MEANINGFUL URBAN
EDUCATION REFORM

SUNY series, Power, Social Identity, and Education
Lois Weis, editor

MEANINGFUL URBAN EDUCATION REFORM

Confronting the Learning Crisis
in
Mathematics and Science

by

Kathryn M. Borman and Associates

Gladis Kersaint

Bridget Cotner

Reginald Lee

Theodore Boydston

Kazuaki Uekawa

Jeffrey D. Kromrey

William Katzenmeyer

M. Yvette Baber

Jessica Barber

State University of New York Press

Published by
State University of New York Press, Albany

© 2005 State University of New York

For information, address State University of New York Press,
90 State Street, Suite 700, Albany, NY 12207

Production by Marilyn P. Semerad
Marketing by Susan M. Petrie

Library of Congress Cataloging-in-Publication Data

Borman, Kathryn M.
   Meaningful urban education reform : confronting the learning crisis in mathematics and science / by Kathryn M. Borman and associates.
      p. cm.— (SUNY series, power, social identity, and education)
   Includes bibliographical references and index.
   ISBN 0-7914-6329-X (hc. : alk. paper) — ISBN 0-7914-6330-3 (pbk. : alk. paper)
     1. Mathematics—Study and teaching—United States. 2. Science—Study and teaching—United States. 3. Curriculum change—United States.
I. Title. II. Series.

   QA13.B67 2005
   510'.71'073—dc22

                                                     2004045256

10 9 8 7 6 5 4 3 2 1

Contents

Illustrations

Tables

Preface

Traditions in both applied anthropology and applied sociology emphasize the importance of the role of the activist researcher. In general, this role links the work done in the field with praxis. In the case of the project undertaken here, there was obvious overlap between the goals of the National Science Foundation's Urban Systemic Initiatives program and an activist/researcher approach to evaluation. Specifically, the mission of the David C. Anchin Center, the institution at the University of South Florida housing the project, is to provide an orientation to school, teachers, and school districts emphasizing the preparation of professionals to meet the needs of children and society in the twenty-first century.

The David C. Anchin Center commits its resources to working with teachers, school administrators, and the community to attain the level of excellence in schools to which every dedicated educator and thoughtful citizen already aspires. In addition to a commitment to staff development that takes teachers' value systems into account, activities in the Anchin Center revolve around concerns about the lives of students and their engagement in high-powered, standards-based, subject matter learning. The Anchin Center's work also emphasizes the importance of the expectations placed on teachers in carrying out systemic reform and the importance of community-based stakeholders. Our work has made us aware of how school culture either sabotages or facilitates change and, at the school and classroom levels, how school district policies and school administrators influence the process of systemic reform, and, finally how student learning in mathematics and science are facilitated through a constructivist pedagogy that requires students to be engaged actively in the learning process.

Acknowledgments

The support and guidance of colleagues and friends throughout the course of our project, "Assessing the Impact of the National Science Foundation's Urban Systemic Initiative," helped contribute to the success of this research. We would like to thank the directors of the Urban Systemic Initiatives and their staff members in our four research sites. Their continuous support of our project enabled us to conduct our research with considerable ease. We would like to extend our gratitude to our Technical Advisory Network members who provided valuable insight during the formative stages of our research as well as throughout the research process. We would like to thank our program director, Bernice Anderson, her staff, and all our colleagues at NSF who encouraged us throughout the life of this grant. We would also like to thank our consultants: Rolf Blank, John Smithson, Ken Frank, Carol Kelley, and those who worked on the pilot community study. Our heartfelt gratitude goes to more than 7,000 people who were participants in our research. They not only provided the data that enabled us to complete our study, but also, more importantly, helped us to understand what occurs in districts, schools, classrooms, and communities involved in systemic reform.

The authors acknowledge the support of the National Science Foundation through NSF Grant # 9874246: "Assessing the Impact of the

National Science Foundation's Urban Systemic Initiative." Any opinions, findings, and conclusions or recommendations expressed in this material are those of the authors and do not necessarily reflect the views of the National Science Foundation.

In addition to the authors of this book, the efforts of many people have contributed directly to this research. These include the following: Timothy Carey, Nikhil Mehta, Ellen Puccia, Karen Moriarty, Susan Greenbaum, Linda Callejas, Julian Smothers, Shantanu Vaidya, Christiana Schuman, Danielle O'Connor, Jessica Pearlman, Troy Sadler, Sandra Cade, Jonathan Gayles, Cathleen Larrimore, Amy Fox-McNulty, Kathleen Del Monte, Jonathan Poehlman, Charmaine Evans, Sandhya Ganapathy, Erin Healy, Ellen Kang, Joy Stallard, and Eric Eyles.

Finally, the authors would like to thank the SUNY press team including Lois Weis, Editor of the Series, Prescilla Ross, Director, Marilyn P. Semerad, Production Manager, and Freelancer Nancy Dziedzic who did a superb edit of the final manuscript.

Abbreviations

CCSSO	Council of Chief State School Officers
Collaborative	El Paso Collaborative for Academic Excellence
CPS	Chicago Public Schools
CSI	Chicago Systemic Initiative
D1–6	NSF Drivers—Driver 1 (D1) Standards-based curriculum and instruction aligned with assessments
	Driver 2 (D2) Unified set of Policies
	Driver 3 (D3) Unified Application of Resources
	Driver 4 (D4) Mobilization of Stakeholders
	Driver 5 (D5) Student achievement in mathematics and science
	Driver 6 (D6) Achievement gap between underserved and nonminority students
D7	Hypothesized seventh driver involving school culture
EPISO	El Paso Interreligious Sponsoring Organization
ESM	Experience Sampling Method
FCAT	Florida Comprehensive Achievement Test
IBEC	Illinois Business Education Coalition
IGAP	Illinois Goals Assessment Program
ISD	Independent School District

LEP	Limited English Proficient
LSC	Local School Council
MUSI	Memphis Urban Systemic Initiative
NAEP	National Assessment of Educational Progress
NCTM	National Council of Teachers of Mathematics
NEA	National Educational Association
NISE	National Institute for Science Education
NSES	National Science Educational Standards
NSF	National Science Foundation
NSLSASD	National School-Level State Assessment Score Database
USI	NSF's Urban Systemic Initiative
PD	Professional Development
PTA	Parent Teacher Association
SASS	School and Staffing Survey by National
SCP	Survey of Classroom Practices
SCQS	School Culture Quality Survey
SEC	Survey of Enacted Curriculum
SIP	School Improvement Plan
SSI	Statewide Systemic Initiative
STAM	Secondary Teacher Analysis Matrix
TAAS	Texas Assessment of Academic Skills
TCAP	Tennessee Comprehensive Assessment Program
TIMSS	Third International Mathematics and Science Study
UTEP	University of Texas at El Paso

1

Historical Context: How National Science Foundation Reforms Build on Earlier Reforms

Urban schools have been the target of a series of reforms over the last twenty or more years. During this time, strategies for school improvement to prevent student failure have addressed a wide range of matters: school safety, computer use, parent involvement, business partnerships, and more. Many claim that these earlier efforts have been piecemeal, targeting one or two dimensions of the schooling process rather than the system as a whole. The National Science Foundation's (NSF) approach, in contrast to the plans that preceded it, emphasizes the simultaneous application of a number of policy levers (called drivers) to move the reform agenda forward. NSF also emphasizes the point that all children can achieve to high academic standards.

In this book we argue that implementing and sustaining systemic reform in mathematics and science for all students requires unremitting effort throughout the national system. Increasingly, with the passage of No Child Left Behind (2001) and related legislation during the George W. Bush administration, education policy emanates from national, state, and local levels. In addition, the federal government has tied funding through national programs such as the reauthorization of the Elementary and Secondary School Act to requirements affecting teaching and learning at the school and classroom levels such as the mandate that all public school teachers will teach only those subjects for which they have been certified. Nonetheless, because the United States is so vast, educational change is still most effectively carried out at the local district level. From the superintendent's office to the classroom, improving mathematics and science practice to benefit all demands that teaching practices are student-centered, taking student needs into account but also assuming all students can and will learn to high standards. Intensive professional development carried out in the school and focused upon both substantive academic content and pedagogical practices is an essential condition for the improvement of student outcomes. In addition, community resources including both parent involvement and the contributions of civic organizations such as museums and local businesses must combine to accelerate student growth and attainment.

Throughout this volume we maintain that unless the strategy in question is designed to (a) promote the academic achievement of all students and to close the achievement gap, (b) engage teachers and school principals in forging a community of learners, and (c) involve parents and stakeholder groups, it will not be sustainable. In agreement with Cuban (2001), we further believe that:

- All schools do not need school-based reform; those schools that do are the lowest performers academically and are the hardest to turn around.
- School-based reform and district reform in urban areas should work in tandem for desired changes to take root and endure at the school site.
- Advocates of research-based school reform who are dependent upon practitioners adhering narrowly to a single design and the use of norm-referenced measures to determine success offer a narrow, technical version of what is a "good" urban school. In doing so, promoters reject teacher expertise while largely ignoring the school's political and cultural dynamics essential for improvement to occur.
- School-based reformers must do their homework on past successes and failures of urban school reform to understand the deeper complexities of the work in which they are engaged (pp. 1–2).

In the remainder of this chapter we turn to an examination of systemwide reform as a national strategy for improving teaching and learning, including a discussion of the NSF driver model as the organizing rubric for mobilizing reform in mathematics and science.

SYSTEMIC REFORM

Although policies at the national level directed toward educational reform predated the 1980s, several prominent documents published in that decade (for example, National Commission on Excellence in Education, 1983; Southern Regional Educational Board, 1981), as well as results from national and international assessments (Crosswhite, Dossey, Swafford, McKnight, and Cooney, 1985; McKnight et al., 1987) revealed the difficulties in providing all students pathways to high-level academic achievement in America's public schools. Most notably, the National Commission on Excellence in Education's report, *A Nation at Risk* (1983), illuminated problems confronting the nation's public education and focused society's attention on the need for a national strategy to improve education. In addition to preserving the status of the Department of Education as a cabinet-level entity, the report also triggered a national debate about what students should be learning and how student learning should be assessed—a debate that is far from over.

According to the report, those working in America's schools were content with low expectations and low standards, and the nation as a whole was in danger of losing its preeminent status as a world power to nations such as Germany and Japan, countries making enormous technological strides at the time. In addition to making a full-scale indictment of the system, the report's authors at the same time called for action to combat "complacency" and "mediocrity." Recommendations included higher standards for students and teachers and an academically challenging core curriculum for all students. Documents demanding reform, coupled with an outcry from business and community leaders, manifest in the Secretary's Commission on Achieving Necessary Skills (1991) report subsequently resulted in policies requiring higher standards for student learning in America's schools. Although the policy-making community still lacked research documenting poor pedagogical practice at the classroom level, there was reason to believe that teacher-centered instruction focused on student learning of facts, basic skills, and not much more was widespread. In fact, weak U.S. student performance on tests measuring the capacity to perform at high levels in comparison with student performance in other nations was the only proof necessary to stimulate the press for reform (Smith and O'Day, 1991).

The current thrust for systemwide reform addresses the assumption that schools have not provided students, especially those students attending the poorest performing schools, with knowledge necessary to be successful in society—the outcome stressed by Newmann (1996) and his colleagues. Students were not sufficiently challenged by the instruction they received, with the result that many were ill-prepared to attend college upon graduation, to enter technologically complex careers, or to engage in challenging intellectual work. By setting more rigorous standards for students, Roeber (1999) and others contended, the general level of student achievement would rise, better preparing students for post-secondary educational opportunities and employment.

The systemic reform movement differs from reform efforts in the past in emphasizing rigorous academic coursework for all students. In addition to improving the overall quality of education for all children, as emphasized in the effective schools approach, educational equity is at the core (Kahle, 1998; Smith and O'Day, 1991; O'Day and Smith, 1993). As a result of this emphasis, the achievement gap between majority group students and students of color is expected to decrease (Williams, 1996). Unlike prior reform efforts, systemic reform recognizes that attempts to change one aspect of the system will require changes in other aspects at all levels of the system. However, the most important change must occur at the level of the school classroom, translated by the teacher and buttressed by policies at the school district and state levels, including those policies and resulting strategies focusing on curriculum, instruction, and assessment.

Led by the National Council of Teachers of Mathematics (NCTM), national content-specific professional groups developed standards to communicate what students should know and be able to do in mathematics (NCTM, 1989, 2000), science (American Association for the Advancement of Science, 1993; National Research Council, 1996), and other subject areas. As described in these documents, instruction should emphasize active learning and higher-order thinking skills, while providing investigative and problem-solving opportunities for all students. Since their introduction, these documents have been used to guide and develop policy at the national, state, and school district levels, in the form of curriculum changes and related high-stakes accountability measures.

The NSF has been critical in providing both the support and the conceptual rubric for institutionalizing systemwide reforms. According to the NSF,

> Systemic reform occurs when all essential features of schools and school systems are engaged and operating in concert; when policy is aligned with a clear set of goals and standards; when forthcoming improvements and innovations become intrinsic parts of the ongoing educational system for all children; and when the changes become part of the school system's operating budget. (NSF, 2000)

Initially, NSF provided support to twenty-five states through its Statewide Systemic Initiatives program (SSI). The goal of the SSI is to assist states in developing the capacity to move from independently devised science and mathematics educational reform measures to state-developed efforts and in coordinating such improvements as teacher preparation, the development of standards-driven instructional materials, and the assessment of student performance. Although large numbers of teachers received curricular and instructional materials, and schools of education undertook some degree of curriculum change in teacher preparation programs, these programs failed to reach those urban schoolchildren in the most difficult circumstances.

To address this need, NSF established the Urban Systemic Initiatives (USI) program in 1993. Funding under the USI program was made available to the urban school districts in the United States with the highest rates of poverty among their school-aged children according to the 1990 census. Twenty-one of the eligible school districts successfully applied for and received a total of 15 million dollars each over a four-year period to carry out systemic reforms in mathematics and science (Westat*McKenzie Consortium, October 1998, p. 6) (see Table 1.1). The funding provided by NSF was considered a medium for developing, expanding, or sustaining reform through partnerships with businesses, educational institutions, and community organizations.

Although NSF did not prescribe approaches to systemic reform, four years into the initiative, and mindful of contextual differences among the sites, NSF developed a model of systemic reform that identified a set of six policy levers, or drivers. Thereafter, grantees were required to assess progress against these drivers (Kahle, 1998). The six-driver model (see Table 1.2) is an interrelated and overlapping structure of

Table 1.1.
National Science Foundation USI Sites

Cohort 1 (Granted in 1993)	Cohort 2 (Granted in 1995)	Cohort 3 (Granted in 1996)	Cohort 4 (Granted in 1998)
Baltimore	Cleveland	Milwaukee	Atlanta
Chicago	Columbus	San Antonio	Jacksonville
Cincinnati	Fresno	San Diego	
Dallas	Los Angeles	St. Louis	
Detroit	Memphis		
El Paso	New Orleans		
Miami-Dade	Philadelphia		
New York City			

Table 1.2.
NSF's Six-Driver Model

Driver	Description
Driver 1	Implementation of a comprehensive, standards-based curriculum and/or instructional materials that are aligned with instruction and assessment available to every student served by the system and its partners.
Driver 2	Development of a coherent, consistent set of policies that support provisions of broad-based reform of mathematics and science at the K–12 level.
Driver 3	Convergence of all resources that are designed for or that reasonably could be used to support science and mathematics education—fiscal, intellectual, materials—both in formal and informal education settings, into focused program that upgrades and continually improves the educational program in science and mathematics for all students.
Driver 4	Broad-based support from parents, policymakers, institutions of higher education, business and industry, foundations, and other segments of the community for the goals and collective value of the program that is based on an understanding of the ideas behind the program and knowledge of its strengths and weakness.
Driver 5	Accumulation of broad and deep array of evidence that the program is enhancing student achievement through a set of indices. In the specific instance of student achievement test scores, awardees on an annual basis are excepted to report the results of student mathematics and science achievement in a multigrade level context for the USI-impacted schools/districts/state(s) relative to appropriate cohort entities (non-USI districts, the state), all of which are defined by the performance baselines.
Driver 6	Improvement in the achievement of all students, including those historically underserved, as evidenced by progressive increments in student performance characterized by the requisite specificity of the USI as a catalytic resource and the appropriateness of attendant attributions.

Note: NSF, 2000.

process and outcome reform drivers. The first four, called process drivers, focus on sustainable success in changing the system's approach to the teaching and learning of mathematics and science. Driver 1 (D1) encompasses the overall intent of the reform effort by encouraging the use of standards-based curriculum that is aligned with instruction and assessment practices. Drivers 2 (D2) through 4 (D3 and D4) support the achievement of the goals of D1. The outcome drivers, Drivers 5 (D5) and 6 (D6), focus on the end result of the reform effort, increasing achievement in mathematics and science and closing the existing achievement gap between advantaged and disadvantaged students. With these drivers in mind, USI programs were designed to be strongly student achievement outcome oriented (D5, D6), with explicit emphasis on resource convergence (D3), establishing a leadership nexus (D2, D4), partnerships that entail more than the provision of resources (D4), and the advocacy role of community relations (D4, D2), in addition to standards-driven curriculum, instruction, and assessment practices (D1). As a result, most projects engaged in a process of policy alignment and employed programs of intensive professional development focused on both subject matter content and constructivist approaches to teaching mathematics and science across grade levels K–12. Other reforms and policy analysts have noted that several policy levers must be engaged in a coordinated fashion. A good recent example is the work of Datnow and her colleagues (2002), whose analysis of comprehensive school reform (CSR) is framed by five "necessary conditions" for success: (1) the success of the reform selection process; (2) quality leadership at site and district levels; (3) CSR design team support and professional development; (4) fiscal resources to support reform; and (5) the reform's ability to help schools meet state accountability mandates.

SCHOOL CULTURE—A POSSIBLE 7TH DRIVER

In addition to policy drivers identified by the NSF, we believe school culture must be viewed as a mediating set of factors that influence the creation of social ties and relationships, and is likely the critical element enhancing or curtailing effective teaching and successful student outcomes. These factors are increasingly seen as vitally important in strengthening or undermining school reform (Fullan, 2001). Additionally, there is evidence to support the notion that student learning is facilitated when principals, teachers, and others develop collaborative relationships within a professional learning community (Newmann and Wehlage, 1995; Louis, Marks, and Kruse 1996; Stein, 1998). According to Supovitz and Turner (2000), teachers who felt supported by their principal reported significantly greater use of reform approaches than did teachers who did not feel encouraged.

Researchers studying the organization of high-achieving urban schools have regarded the development of professionally enriching work groups as a major facilitator of commitment and effort with the potential to improve student learning (Louis, 1998, p. 1). Those who study the social organization of schools also emphasize "the multiple, embedded ways in which teachers seek and use professional networks to increase their knowledge and skills" (Louis, 1998, p. 2). While high-quality instruction depends on the competence and attitudes of individual teachers, individual knowledge, skills, and dispositions when put to use collectively within a school can create a strong professional community (King and Newmann, 2000).

THE CURRENT CRISIS: CLOSING THE PERSISTENT ACHIEVEMENT GAP

The United States requires a literate, technologically savvy, mathematically proficient population capable of high-level problem solving. Students in turn require an environment in which educators can teach the materials effectively and challenge students to think creatively while instilling appropriate knowledge and skills as a foundation for future learning. The first issue to acknowledge is the persistent educational achievement gap between students of color and white students:

> This ethnic educational achievement gap is hardly news. It is a well-studied and well-established fact, using almost any measure (the well-known 15-point average I.Q. gap between blacks and whites sensationalized by *The Bell Curve*, SAT scores, college and high school grade-point averages, graduate and dropout rates) black students nationwide do not perform as well as whites. (Singham, 1998, p. 8)

Corresponding gaps are also evident in financial inequities that persist in school funding; in schools that cannot keep up with high technology in either equipment or teaching; in tracking and course offerings that offer some students more challenging academic content and leave others with courses such as "basic math" or "general science." Although the achievement gap is well documented in the United States, no clear conclusion has been reached to explain the persistence of the gap, and many "measures" showing disparities in cognitive abilities between groups have been discredited as valid indicators of achievement differences.

Making meaningful and durable systems change requires both taking into account the unique nature of each school and community and applying standards-based criteria to evaluating teaching, learning, and other features. Change must be structured to be adaptable to different populations and regional areas without sacrificing an adherence to the goals of high achievement and continued learning of all students. Together with standards for student achievement, standards for teaching are important because teacher knowledge is a critical component in the classroom (Haycock, 2001). The importance of these factors is reflected in the USI reform efforts through increased standards-based instruction and professional development for teachers in urban school districts. By increasing policy makers' and the public's understanding of the varying impacts of systemic change at the state, district, and school levels, systemic reform efforts including the USI offer multiple avenues for educational improvement and the reduction of achievement differentials (Corcoran and Goertz, 1995).

NSF-sponsored USI reform efforts attempt to reduce the achievement gap and address factors that contribute to it. Research has identified several problems faced by teachers, students, and administrators in urban school districts and schools. First, low-performing schools often suffer from a lack of coherence among activities deemed as priorities by school district or school-level administrators (Bryk, Sebring, Kerbow, Rollow, and Easton, 1998). The lack of coherence includes fragmentation of the curriculum, fragmentation or lack of coordination in organizing the school day, poorly related or incompatible instructional strategies, inconsistent behavioral expectations, and the lack of a shared purpose and shared values. Second, many of these schools face a poorly organized or nonexistent program to support the acquisition of knowledge and skills needed to engage in effective reform at the school level. In addition, teachers and principals lack the time and resources to mobilize the information, skills, and knowledge to undertake radical transformations of teaching, learning, and assessment, as well as school organizational change. Third, there are major disincentives for teachers to elect to work in schools with histories of failure much less to stay with schools undergoing major shifts in practice. Developing pedagogical skills, the capacity to undertake complex organizational transformation, and the willingness to stay with a career with minimal financial rewards for the work required seems a most difficult challenge for even the most dedicated teachers and school staff.

Measures such as standardized testing, increased school accountability, community involvement, and resource allocation must also be structured and evaluated in ways that relate them to achievement differentials. An increasingly important factor is the use of technology as part of school and work environments. When business leaders began to worry about applicants meeting their minimum literacy requirements and possessing higher-level thinking skills, the question of educational efficacy came to the forefront (Mizell, 1992). The commitment of educational leaders to increase the achievement levels of all students should be a real part of reform efforts to make certain all students are ready and able either to continue into higher education or to be successful in a rapidly changing, technically demanding job market. Systemic changes in both structure and practice are important to this goal and to the reduction of achievement differentials through increased alignment of teaching philosophy and policies with classroom practices.

In the first part of this chapter our goal has been to set the stage for a discussion of the nature of the work that we undertook in our study of the implementation of NSF's Urban Systemic Initiative program in four sites. Our aim has been to show how a concern with the failure of so many students to attain to high levels of achievement in mathematics and science coupled with the perception by policy makers that most classroom instruction in the nation's public schools was teacher-centered rather than student-centered led to the development of strategies to improve both conditions. In addition, it was widely (and in our view correctly) perceived that reform on the revolutionary scale envisioned by NSF required the concerted efforts of many dedicated individuals throughout the system and school community. Without the targeted efforts of superintendents both to create the vision and to mobilize resources in concert with those of the school principal, teachers, parents, and community members, systemic reform is impossible to achieve. In the next section of the chapter we turn to a discussion of the research plan we followed in building an understanding of the impact of systemwide reform.

ASSESSING THE IMPACT OF THE URBAN SYSTEMIC INITIATIVES

During the course of our three-year project, we researched the effectiveness of NSF's driver model in explaining educational reform and estimated its effectiveness as a template for district and school policies designed to close the achievement gap and increase overall success in science and mathematics. The participating cities in our evaluation project (Chicago, El Paso, Memphis, and Miami-Dade) were chosen because they provided a reasonably representative sample of the twenty-one districts supported by NSF's USI. These four cities vary by region, size, predominant ethnicity, and native language of students and their families. Because we intended to carry out an assessment of both the efficacy of the driver model and the effectiveness of USI programs, our methodological approach is understandably complex. Below we provide a brief description of the methods used in this research.

The major goals of this research study were: (1) to assess the impact of USI reforms by modeling relationships between each latent variable (driver) and a set of manifest indicator variables; (2) to determine how reforms in curriculum and instruction affect teacher and student outcomes at the classroom level through the enacted standards-based curriculum in mathematics and science; and (3) to investigate the roles of leadership, local resources, and national, state, and local policies related to systemic reform that foster or inhibit student achievement outcomes and outcome differences. Our research plan called for three studies (the Mathematics and Science Attainment Study, the Study of the Enacted Curriculum, and the Policy Study) to be conducted over a three-year period simultaneously. These interconnected studies included a number of questions that guided the work:

Mathematics and Science Attainment Study:

- What is the impact of successful initiatives funded through NSF's USI program?
- To what extent do both (a) student achievement outcomes in general and (b) attainment outcomes between underserved students and their peers differ as a function of the duration and intensity of treatment?
- To what extent is the causal structure implied by NSF's reform drivers consistent with the observed relationships among the indicators of each driver?
- What are the similarities and differences in causal structures across the four participating sites?
- What evidence suggests that additional drivers are causally related to achievement in general and to attainment differences specifically?
- What causal structure for interpreting the data provides a consistent, reliable, and cost-effective model for other USI sites to use in carrying out assessments of the impact of their programs on student achievement outcomes?

Study of Enacted Curriculum:

- To what degree are science and mathematics being taught in a manner consistent with local goals, state frameworks, and national mathematics and science education standards?

Policy Study:

- What is the impact on local USI reform of national, state, and local policies related to systemic reform, including assessments, standards, and professional development?
- How are resources mobilized among various constituencies such as schools, universities, business and industry, and political agencies?
- What is the role of institutional and individual leadership in this process?
- What community and district contextual characteristics are particularly important in affecting student outcomes?

A multisite, multilevel case study design was employed to provide perspectives at four levels: the school district, the school, the school's community, and school classroom sites implementing NSF-sponsored initiatives. Our array of studies and research questions allowed us to develop multiple indicators for each of the NSF drivers, to do so over a three-year period, and ultimately to execute a comprehensive analysis of multiple salient features fostering or inhibiting possible student outcomes. These included both organizational and individual level factors such as student achievement and student engagement (D5 and D6); teachers' reports of their professional development experiences, use of technology in the classroom, and involvement in decision-making processes (D1 and D3); school district assessment practices and school-level support structures (D2); teachers' classroom practices (D1); the nature of community-school partnerships and other arrangements with business and industry (D4); school climate and school leadership (D3 and the hypothesized D7). In addition, we attempted to contextualize our work by taking into account factors including local, state, and national programs and the range of important local and regional characteristics.

The case study design incorporated a variety of quantitative and qualitative research methods. The inclusion of multiple methods enhances the rigor of the research as the limitations of one method may be balanced by the strengths of others (Kromrey, Hines, Paul, and Rosselli, 1996). Further, given the complexity of systemic reform, the use of multiple methods increases the likelihood of accurately representing the myriad of influential variables at work to either undermine or sustain reform.

THE SCHOOL DISTRICTS AND CITIES

We worked with USI directors and others at each of four sites to allow comparisons among programs and to estimate the extent to which results may be generalized. Sites vary by region, size, and predominant ethnic composition of students. Size differences are important in view of the relatively greater challenges facing large urban school systems, and because NSF funding at 15 million dollars over four years was the same in all USI-funded locations regardless of number of students served. Racial/ethnic variation addresses the importance of language, culture, and racial or ethnic discrimination as variables affecting educational performance.

At the start of our study, all four sites had operated USI programs for at least four years. Chicago and Miami-Dade are very large districts (ranking third and fourth nationally), while Memphis and El Paso are comparatively smaller. Students in Memphis and Chicago are predominantly African American (70 and 55 percent, respectively), while those in El Paso and Miami-Dade are predominantly Latino (85 and 52 percent, respectively), although of different national origins (i.e., Mexican American and Cuban American). Chicago has a substantial Latino student population (31 percent), whereas Memphis has a relatively small percentage (1 percent). Similarly, Miami-Dade has a substantial proportion of African American students (31 percent), while El Paso has relatively few (3 percent). In various combinations, therefore, all four districts have very high numbers of traditionally underserved students and all reflect high levels of family poverty and chronically low levels of student achievement. All four cities faced state-imposed mandates to improve school performance and to raise test scores of students. Below we provide a brief characterization of each of the sites.

Chicago

While Chicago's economic growth in the 1990s was somewhat less substantial than the prosperity experienced by its U.S. counterparts—New York and Los Angeles—in the global finance economy, Chicago still managed to gain both jobs and population during this period (Abu-Lughod, 1999). Currently, Chicago headquarters more of the nation's largest firms than Los Angeles and has also developed capacity to compete in global financial markets through its venerable institutions, most notably the Chicago Mercantile Exchange. Nonetheless, Chicago remains one of the most segregated cities with respect to both income and race/ethnicity. Abu-Lughod (1999) asserts that despite decades-old open housing laws, the physical isolation of blacks and, increasingly, of Latinos continues:

> Over time . . . racial segregation has persisted and, with the flight of Caucasians either beyond the city limits or into explicitly identified city

"bunker" zones, areas of minority residence have expanded without leading to greater integration. (p. 332)

Chicago's schools, in part because of an emphasis on neighborhood-based magnet programs, exhibit similar patterns of segregation as the neighborhoods in which they are located.

As the nation's third largest district, it is not surprising that Chicago Public Schools (CPS) over the past thirty years has alternately served as a model for exemplary practices and as the case-in-point for mismanagement, chaos, and abysmally poor student performance (Hess, 1991). At present, the school district's 596 elementary and high schools serve nearly 435,476 students, an increase of more than 10,000 students since 1999. In the early 1990s CPS was an ideal candidate for NSF's USI, with low student achievement, high poverty, and a diverse but racially isolated student population. During the 1990s and into the twenty-first century, both the city and the city's schools have reversed disturbing earlier patterns of population loss, with a particularly notable growth in its Latino population and a decline in numbers of youth ages five to seventeen living below federal poverty levels. This population declined from a high in 1990 of 190,000 (21.4 percent of that population) to 182,000 or 18.5 percent of youth in this age range.

The former CEO of CPS, Paul Vallas (now Superintendent for Philadelphia Public Schools), served as the school district's leader during the 1990s as an appointee of the mayor and attributes the district's success in maintaining enrollments of both low- and upper-income students to the press for high achievement for all students.

Educational Reform in Chicago. Chicago's history of educational reform is closely tied to the corporate sector, the mayor's office, local community activists, and the state legislature. The contrasts in allocation of power and influence over educational policy in the city are dramatically illustrated by the contrasting policies of two reform laws—one passed in 1988 establishing Local School Councils (LSCs) in each of the city schools and the other the June 1995 reform law enacted by state legislation, centralizing control in the mayor's office (Shipp, 1998). The latter state law concentrates management of the districts in the hands of a chief executive officer and board of trustees, all of whom are appointed by the mayor. While LSCs remain nominally in charge of budgetary priorities at the school level and retain the power to hire and fire school personnel, the more recent policies give ultimate control for systemwide initiatives and accountability systems to a centralized management team.

Yet it wasn't always so. In the 1980s, as the city's school bureaucracy grew, parents, community members, and the city's business establishment formed a coalition linked by their strong belief that bureaucratic bloat and central office decision making were both fiscally irresponsible and highly detrimental to students' academic success (Bryk, Sebring, Kerbow, Rollow, and Easton, 1998). LSCs were created to channel community support to the school. It was believed that strengthening school autonomy and control and establishing democratic local governance through the LSC would enhance school performance. This approach was self-consciously aligned with an effective schools research agenda arguing for both site-based decision making and parental involvement as the keys to improved student achievement. Schools continue to be governed by LSCs staffed by six parent representatives, two community representatives, two teachers, one principal, and, in high schools, one student representative. However, by the mid-1990s many in the corporate sector determined that decentralization must be balanced by stronger control over the system's resources to insure success for all students.

During the early 1990s business leaders across the state created a new coalition that included the Illinois Manufacturers Association, the Illinois Business Roundtable, the Illinois State Chamber of Commerce, and the Illinois Retail Merchants Association in league with the Chicago Civic Federation and the Chicagoland Chamber of Commerce under the umbrella of the Illinois Business Education Coalition (IBEC) to improve accountability (Shipp, 1998). Together, according to Shipp:

> All could agree on stricter accountability of principals to the district and on central office leadership modeled after an up-to-date corporation as well as on reinstating unfettered mayoral discretion in the selection of the school board. (p. 177)

While the mayor received no additional funding under the reform legislation of 1995, he did enjoy considerable control of both district-wide policy initiatives and funding allocations. During the same year, the mayor appointed a new team of managers whose goal of improved classroom performance included a spectrum of school intervention programs and teacher accountability measures to improve academic instruction and adherence to standards.

Chicago Urban Systemic Initiative (CSI). The CSI builds upon and aligns management accountability in compliance with state and federal education standards. In addition, CSI was designed to be the organizing framework for all mathematics, science, and technology reform efforts within the Chicago Public Schools. CSI's goals include improving the scientific and mathematical literacy of all students in a technological society and motivating students to pursue careers in mathematics, science, engineering, and technology. Since its inception, CSI has implemented programs and assessment tools to produce major systemwide reform in mathematics, science, and technology teaching and learning. Not unlike the implementation of USI programs in other sites, CSI program developers initiated their efforts by conducting a self-study of district mathematics and science instructional programs. The 300 schools participating in the initiative represent about half of all CPS schools. Nonparticipating schools were, nonetheless, affected through the emphasis systemwide on standards-based criteria.

El Paso

El Paso is located on the Mexican border, with a population approaching one million, and is the fourth largest city in Texas. When El Paso's population numbers are added to the 1.3 million persons residing in the sister city of Ciudad Juarez, Chihuahua, Mexico, the El Paso-Juarez area ranks as the largest binational community on any international border anywhere in the world (Anderson, 1997). The vast majority of students attending school in the three largest districts— El Paso, Ysleta, and Socorro Independent School Districts—are from low-income families and close to half of the children enter school with limited proficiency in English (Navarro and Natalicio, 1999). El Paso ranks as the fifth-poorest congressional district in the United States. Not surprisingly, educational issues in El Paso include bilingual and immigrant education, as well as professional development and teacher education, improving partnerships with parents, family, community, and businesses, and accentuating the role of leaders in participating school districts and schools. These issues become increasingly important when considered in the context of El Paso's rapid population increase in the 1990s.

The overall population in El Paso has grown by more than 8 percent during the past ten years. El Paso's youth ages five to seventeen living below federal poverty levels decreased only slightly from 48,188 to 48,661, a decline representing a change from 35 percent of this age group to 31 percent. El Paso's jobs in distribution, transportation, and the service sector represent opportunities for individuals struggling in the rural areas of the borderland, providing upward social and economic mobility for families moving across the border from Ciudad Juarez. Many of these adults have been educated primarily in Spanish, but their children find themselves in English-speaking classrooms, challenging El Paso to develop a multilingual solution to the problem of increasing student achievement. We were impressed throughout our work in El Paso by how schools in each of the three districts we studied (El Paso, Socorro, and Ysleta) embraced students who were Mexican citizens making the trip across the city's bridges daily to attend school in the United States.

Educational Reform in El Paso. Educational reform in El Paso has in many respects paralleled educational reform in Texas. School finance reform following an important court case (*Edgewood v. Kirby*) in 1989 preceded reforms in the 1990s aimed at increasing the achievement of all students and strengthening school accountability across the state. Following *Edgewood v. Kirby*, a more equitable system of district financing was established, resulting in improved conditions in schools throughout Texas attended by Latino and black children. In 1990 the state legislature passed a law mandating high-stakes assessment in the form of the Texas Assessment of Academic Skills (TAAS). The TAAS measures student achievement in the areas of reading, writing, mathematics, social studies, and science. In 1995 the State Board of Education adopted exit test requirements for students' graduation eligibility.

In El Paso a growing realization that existing reforms were not increasing the participation of all students in high-level courses, college and university enrollment, and graduation and other attainments prompted community leaders to create the El Paso Collaborative for Academic Excellence (Collaborative). In early 1992 the president of the University of Texas at El Paso brought together key stakeholders from El Paso's business and educational sectors. These leaders included the superintendents of the three largest school districts (El Paso Independent School District [ISD], Ysleta ISD, and Socorro ISD), the president of El Paso Community College, the heads of the two chambers of commerce, the mayor and the county judge, and the lead organizer of the El Paso Interreligious Sponsoring Organization (EPISO), a state-funded, faith-based organization, arguably the most active and effective grassroots community organization in the region. Together, they launched the Collaborative, a community-wide systemic reform effort aimed at improving the academic achievement of all students in El Paso, K–16.

Each school district participating in the Collaborative's work has continued to frame its own agenda for educational reform. For example, the Ysleta ISD, regarded as progressive and innovative, addressed the ninth grade dropout problem through a program of "small schools." In each of the district high schools teams of teachers work with groups of twenty-five freshmen that attend classes in areas of the building apart from upper classmen. In a recent year the Ysleta ISD reported dramatic results: of the 3,407 freshmen enrolled in district schools in 1999, 3,305 or 97 percent earned sufficient credits to become sophomores the following fall, compared to the Texas average of 82 percent (*El Paso Times*, 9/27/99). In the El Paso ISD parents hired as full-time employees serve as liaisons between the community and the school. El Paso ISD has seen a stunning reversal in dropout rates through an emphasis on close contact and careful monitoring by these members of the community. To develop its dropout programs, each district worked with faculty colleagues at the University of Texas at El Paso (UTEP). UTEP's visibility in the El Paso community has also been critically important to the Collaborative's work in implementing the El Paso USI.

El Paso Urban Systemic Initiative. The NSF's USI grant enabled the Collaborative to carry forward its commitment for systemic reform in the three largest school districts. During the summer of 1992 the Collaborative began a program of professional development activities with the establishment of the Teams Leadership Institute. The Institute brought together leadership teams for three days from the forty schools that volunteered to participate in the Collaborative. To qualify for participation, schools were required to meet conditions likely to sustain reform activities including carrying out initial assessments to determine patterns of student achievement and failure. Teams from each school were composed of seven to ten professionals including the principal, assistant principal, a counselor, and five to seven teachers. After assessing school needs, teams drafted preliminary action plans and methods for sharing information with school staff. Follow-up institutes were provided for three years. In addition, two-day institutes were held for principals and subject-matter institutes were held for teams of five to seven teachers from participating schools that also met for eight additional days during the school year. The Collaborative also provided on-site mentoring with experienced teachers sent to assist teachers in their classrooms as they introduce new curricula and make changes in their teaching practices.

Because the Collaborative's work is administered though offices at UTEP, the three districts involved have been spared the problem of creating administrative structures for program delivery. Nonetheless, ranking district administrators in at least one of the districts, Ysleta ISD, see arrangements with UTEP as another layer in a bureaucratic system impeding the smooth and coordinated delivery of aligned programs in mathematics and science at the classroom level. It can also be argued, however, that in a context of unstable and volatile school district politics—all but one of the districts during the course of the initiative fired, suspended, or brought suit against an incumbent superintendent—the role of the Collaborative as an independent third party has been critical in creating stability and consistency in the administration of the USI program.

Memphis

Memphis is Tennessee's largest metropolis and the eighteenth largest city in the United States. Memphis has the country's ninth-largest African American population, representing nearly 50 percent of the city's inhabitants (*Heritage Memphis Culture Guide*, p. 5). Overlooking the Mississippi River, Memphis is a major port, rail terminus, and marketing center serving the South's lumber, livestock, and cotton industries. Despite the economic decline in recent years, Memphis remains the capital of the mid-South and continues to be a magnet for economically disenfranchised people who leave their homes in the rural areas of Tennessee, Mississippi, and Arkansas.

Educational Reform in Memphis. Superintendent Gerry House implemented a multitude of reforms, including the USI, and spearheaded educational reform in Memphis from 1994 to 2000. The original reform agenda articulated in the Memphis proposal submitted to NSF to qualify for USI funding addressed the needs of both African American and white students. However, the district during this period witnessed the continued departure of white families (and students) to county school systems and the arrival of individuals and families of Latino and Asian descent. Individuals of Asian or Latino origin increased from 1.4 percent to 4.5 percent of the total population in the 1990s, and the increase is projected to continue. The students from these families bring the challenge of linguistic difference to a district that has not been focused on cultural and language variation until now. Since the mid-1990s conditions have improved for the staff and students of Memphis City Schools. The system is one of the few in the state fully accredited by the Southern Association of Colleges and Schools. Additionally, fewer children ages five to seventeen live in poverty, that number decreasing from close to 40,000 (representing 25 percent of children in this age category) to just under 39,000 or 21 percent of the overall population.

The year 2000 saw the resignation of the reform superintendent, the selection of a new superintendent who is native to the city and school district, and the initiation of a new conversation about the district's future. The departure of Superintendent House signaled a more measured approach to reform and a dramatic shift in emphasis. The new superintendent and native Memphian Johnnie Watson conducted what some viewed as a poorly designed study of the impact of school reform models on student achievement and concluded that student performance had in some cases actually declined but at best had remained unchanged (Ross, personal communication, July 19, 2001). Beset with concerns about the large number of city schools making Tennessee's list of failing schools (twenty-six of the state's forty-eight), Superintendent Watson had little choice from his perspective but to target teaching and learning in new ways (Ross, 2001, August 19). His goals included the adoption of a districtwide curriculum to insure a similar approach to instruction in the district's schools. Although this strategy suggested a retreat from more creative and innovative approaches, it was also a hedge against negative impacts on highly mobile student populations in Memphis schools.

The Memphis Urban Systemic Initiative. The Memphis Urban Systemic Initiative (MUSI) decentralized the system's organizational structure from four subdistricts to twelve clusters containing twelve to fifteen schools coordinated by an instructional administrator reporting to the USI director. A major focus of MUSI was the elimination of all low-level science and mathematics courses, beginning at the high school level, and moving progressively down through the lower grades. Another focus was professional development for all teachers in all schools. According to Memphis reports, a total of 112,000 (100 percent) of elementary, middle, and high school students were learning mathematics and science in classrooms that were directly supported by the MUSI.

Miami-Dade County

Miami, Florida, is a city that has captured the imagination and interest of politicians, entertainers, investors, and others during much of the twentieth century, an interest that intensified in the 1990s. During this period, immigrants from Latin America and the Caribbean who settled in Miami transformed the city. The schools as well as political and economic institutions have been dramatically affected (Stepick, Grenier, Castro, and Dunn, 2003).

Immigrant power, according to Stepick and his colleagues (2003), is a reality in Miami where "Latinos' political presence combined with their economic and cultural ascendancy challenge the corporate leaders' earlier hegemony in setting the local agenda" (p. 35). By the year 2000 Miami's Latino population had "made Miami the economic and transportation gateway of the Americas" (p. 20) despite having only 5 percent of the nation's total Latino population. In fact, more than 40 percent of the largest Latino-owned businesses in the U.S. are located here.

In the wake of the 2000 national elections the U.S. president and political parties in general are reaching out to Latinos, especially those in Florida, the site of political angst and chaos in the presidential contest. This process is pervasive throughout the nation. Latino candidates ran for mayor in many of the nation's major cities in 2001, and, although they lost in the largest three (New York, Los Angeles, and Houston), this political activity sets the stage for a new generation of Latino municipal leadership. Latino candidates also campaigned for statewide offices in several states in the 2002 elections. This new centrality of Latinos to national, state, and local politics has both substantive and symbolic dimensions. Even at the symbolic level, the outreach represents a significant improvement over the neglect that Latinos have experienced in most previous elections. The current political power of Latinos is critical in understanding Miami-Dade's recent history as well as the histories of El Paso and Chicago.

The Miami-Dade County public school system is the fourth-largest school system in the nation and the largest countywide district in Florida. Miami-Dade County is home to a culturally mixed population, of which two-thirds of its citizens are of Cuban heritage. While individuals enumerated in the 1990 census came primarily from Cuba and other Spanish-speaking islands in the Caribbean, new residents hail from Brazil, Argentina, Honduras, and other countries in South and Central America. These newcomers bring diversity in cultures and lifestyles, as well as a diversity of prior educational experiences. In the 2000 census the city of Miami reported the largest percentage (61 percent) of nonnative residents of any U.S. city. In addition, Miami-Dade, like Chicago, presents a complex economic landscape with highly impoverished neighborhoods located city blocks from expensive high-rise condominiums. Although the relative size of the population of youth ages five to seventeen living below federal poverty levels actually grew from 75,703 to 91,715 between 1990 and 2000, this number actually represented a decline from 23.6 percent of that population to 22.7 percent.

A defining moment for the city of Miami came in 1959 with Fidel Castro's takeover of Cuba. Opposition to Castro's transformation of the island nation into a Marxist state led to an exodus of Cubans to Miami-Dade beginning in 1965 with U.S.-sponsored "Freedom Flights." During the spring of 1980 the Cuban Mariel Boatlift carried 125,000 refugees to the south end of Miami Beach seeking asylum and a new life, and also ushered in an unprecedented influx of refugees from Haiti, Nicaragua, and elsewhere in the Caribbean. By the end of the 1980s the Cuban refugee population dominated the city's political and social landscape. During the 1990s Miami-Dade's large Cuban population—whose countywide numbers exceeded 600,000—was actively engaged in the political process, holding dominant positions in the city government and school district board, showing the demographic changes in Miami-Dade since the 1990 census.

Educational Reform in Miami-Dade. In 1986 Miami-Dade County Public Schools implemented a plan for school-based management and shared decision making that gave local schools increased control over their own budgeting, resource allocation, and curriculum. This restructuring was similar to efforts in site-based decision making that took place in Chicago at the same time. Indeed, structural changes in 1996 reshaped school board elections and patterns of representation. Single-member district elections enhanced the representation of African Americans on the school board and fostered a decentralization movement leading to both central office staff downsizing and school site administrator power enhancement. However, by 1989, it became clear that these decentralization reform efforts in Miami-Dade affected organizational structure much more than instructional processes. Student achievement, school completion, and attendance remained low (Rungeling and Glover, 1991).

Miami-Dade County Urban Systemic Initiative. In 1993 Miami-Dade County Public Schools applied for NSF's Urban Systemic Initiative funding, planning to involve every school and all district students in improving mathematics and science achievement by the end of the third year of the initiative. Using NSF funding in conjunction with other federal resources (for example, Title II Eisenhower mathematics and science funds), the Miami-Dade USI aimed to strengthen content and instruction, enhance student learning, coordinate resources, and establish equitable practices in school mathematics and science. It also intended to make connections with community partners to link mathematics and science learning to real-world issues and careers.

School Selection

Within each site, elementary, middle, and high schools were selected in a three-step process, in consultation with our site coordinators, our Technical Advisory Network,[1] and USI project staff at the sites. The goal of the school sampling process was to obtain samples of schools that: (1) were reasonably representative of the school district; (2) evidenced substantial variability in the extent to which the schools participated in the USI reform initiatives; and (3) were likely to present limited confounding of driver variability with variability in other, extraneous variables. A total of twenty elementary schools, fifteen middle schools, and twelve high schools across the four cities participated in our research. The forty-seven schools included in our study were located in all parts of the four cities. In Chicago, for example, the six elementary schools were scattered throughout the city's six regions, from the South side, close to Indiana, to Uptown in the north.

Teacher Selection

Five teachers were selected for in-depth study within each of the sampled schools. Initially, we attempted to select teachers who represented subject areas of mathematics and science, who represented a variety of grade levels in each school, and who had at least five years of teaching experience at the sampled school. Ultimately, 230 teachers were selected using the following criteria: (a) teachers who had appropriate certification; (b) teachers in the original sample who had been teaching at the school for more than three years; and (c) teachers who were willing to participate in the study throughout its three-year duration. Additionally, we selected teachers to insure representation by gender and ethnicity in the sample. Finally, the research team selected teachers at grade levels three through five at the elementary level and selected teachers of algebra and biology ("gateway" courses) at the secondary level whenever possible. This was done to insure inclusion of teachers whose students were likely to be taking state- and district-mandated tests. Table 1.3 presents demographic information for participating teachers. These teachers were highly experienced; the majority had twelve or more years of teaching experience. The representation of females (70 percent) to males (30 percent), while unbalanced, reflects the distribution of the teacher population. These teachers were ethnically diverse across the sites and their diversity varied by city. For example, in El Paso, the majority (60 percent) of teachers are Latino.

Data Collection and Analysis

A variety of data collection procedures were employed in this study to obtain information from key individuals involved in the process of systemic reform, including individuals from the community, representatives from the district, school-level leaders, teachers, and students. Methods used to collect these data included classroom observations, interviews, surveys of teachers and students, experience sampling, and archival document review. We used qualitative and quantitative methods to analyze the myriad of data that was collected. Table 1.4 provides the methods of collecting data, the various sources used to collect that data, and the primary modes for analysis. See Appendix A for a detailed description of the instrumentation.

Table 1.3.
Characteristics of Participating Teachers

	Chicago		El Paso		Memphis		Miami-Dade		Total	
	n	%	n	%	n	%	n	%	n	%
Gender										
Male	18	36	19	38	10	16.7	22	31.4	69	30
Female	32	64	31	62	50	83.3	48	68.6	161	70
Total by Site	50		50		60		70		230	
Race/Ethnicity										
Caucasian	18	36	14	28	23	38.3	32	51.4	91	39.6
African American	27	54	3	6	37	61.7	22	31.4	89	38.7
Latino	2	4	31	62	0	0	12	17.4	45	19.6
Asian/Pacific Islander	0	0	1	2	0	0	0	0	1	0.4
Multiracial/Other	3	6	1	2	0	0	0	0	4	1.7
Years teaching										
1–2 Years	0	0	3	6	4	6.7	1	1.4	8	3.5
3–5 Years	2	4	6	12	6	10	9	12.9	23	10.0
6–8 Years	1	2	11	22	7	11.7	9	12.9	28	12.2
9–11 Years	3	6	3	6	3	5.0	6	8.6	15	6.5
12+ Years	32	64	25	50	34	56.7	43	61.4	136	58.3
Missing Data	10	24	2	4	6	10	2	2.9	20	9.6
Years in Present School										
1–2 Years	3	6	14	28	7	11.7	3	4.29	27	11.7
3–5 Years	2	4	12	24	18	30.0	14	20.0	46	20.0
6–8 Years	5	10	8	16	7	11.7	14	20.0	34	14.8
9–11 Years	4	14	3	6	8	13.3	9	12.9	27	11.7
12+ Years	13	26	12	24	15	25.0	29	41.4	69	30.0
Missing Data	20	40	1	2	5	8.3	1	1.4	27	11.7

Table 1.4.
Data Sources and Analysis Procedures

	Sources	*Analysis Procedures*
Documents	NSF performance evaluation reviews, NSF end of year reports, School improvement plans	District and school documents were organized and analyzed using data matrices and coded scoring criteria to inform the Structural Equation Model (SEM).
Interviews, Focus Groups, Question-naries	District level officials, principles community members teachers, students	Interview and focus group data were audio-taped, transcribed, and analyzed long with open-ended items from e-mail correspon-dence with teachers using qualitative data analysis procedures (Miles and Huberman, 1994). Typically, multiple researchers read the data to identify categories, themes, and emergent patterns. In some cases, partici-pant responses were placed on matrices to facilitate cross-case analysis. Some cate-gories identified in this fashion cut across all sites while others were specific to an individual school or research site.
Surveys	Teachers, students	Individual surveys were analyzed using a variety of quantitative data analysis procedures that include cluster analysis, factor analysis, Rasch modeling, and Structural Equation Modeling.
Obser-vation	Classrooms	Observations were analyzed using two instruments: the classroom observation checklist and the authentic instructional practices coding matrix.
		The checklist uses mathematics and science standards, as well as components of other instruments designed to identify five mathematics and science standards-based practices (Sykes, 1999; National Council TM, 1991; NRC, 1996). All observations were coded using the Authentic Instruc-tional Practices Coding Matrix developed using two major sources: Authentic Instruction Classroom Observation Form (Borman, Rachuba, Datnow, Alberg, MacIver, Stringfield and Ross, 2000;

(continued)

Table 1.4.
(continued)

Sources		Analysis Procedures
		D'Agostino, 1996), and the Secondary Teacher Analysis Matrix (Simmons, Emory, Carter and Coker, 1999). This Coding Matrix is composed of four categories: classroom discourse, social support, student engagement, and lesson coherence.
Student Achievement Data	State assessments (TAAS, FCAT)	Quantitative data obtained from the variety of sources described in this chapter were integrated into a multilevel structural equation model. The type of model employed for this analysis represents a combination of traditional structural equation modeling (used for the identification of potential causal relationships among latent variables) and hierarchical linear modeling (used for modeling relationships among nested factors).

RESEARCH ON SYSTEMIC REFORM

The research undertaken in our four urban sites with large numbers of participants at multiple school and grade levels engaged the efforts of a large and diverse team. Our approach emphasized the use of a mixed-methods design and a complex array of surveys, interviews, classroom observations, focus group protocols, and a considerable amount of time in the field in schools, classrooms, district offices, and stakeholder workplaces and homes. The result is that our findings are as rich and complex as the methodological approach we have outlined in this chapter. A major assumption underlying this work is that whole scale reform in science and mathematics requires a program of research that is multifaceted and multilevel. Thus, as shown in Table 1.4, we undertook archival research by analyzing documents including annual school improvement plans and progress reports as well as undertaking interviews with teachers, principals, superintendents, and their staffs, and community-based stakeholders in addition to carrying out extensive observations in countless classrooms in the forty-seven schools included in this study. In the remaining chapters of this volume we tell the story of how systemic reform was implemented in the four sites by considering the multiple levels and actors at each level who had a role in carrying out USI programs in their districts.

2

The Importance of District and School Leadership

This chapter presents findings from our research that address the questions that guided our policy study. Specifically, we address the question: What is the role of institutional and individual leadership in systemic reform? By looking critically at the role played by key players at the district and school levels, we are able to achieve a better understanding of the impact of their involvement in policy making on systemic reform. Interviews with these district- and school-level administrators in each of our sites provided important information about the place of the USI initiative in the district's reform agenda. NSF requires districts to position the reform at the center of all that the district does. The data we gathered through intensive interviews with district staff and principals help us understand the emphasis given to providing resources needed for systemic change, including professional development, school staffing, student grouping arrangements, assessments, and the implementation of challenging curricula in mathematics and science. Although the NSF provided 15 million dollars over four years to each of the urban sites it supported in making systemic change, this amount is relatively small in districts the size of Chicago or Miami-Dade, where annual school budgets range in the billions of dollars. In each of our four sites we found that resource allocations for professional development and curricular materials, including technology such as graphing calculators for students, commanded the bulk of the expenditure to promote systemic changes in mathematics and science.

All districts recognized that resources must be directed toward teachers and classrooms to improve student achievement in mathematics and science. However, along with carrying out systemic reform efforts, district and school administrators must negotiate state educational policies, as well as address local school board policies or politics and school-level concerns. It is no surprise that these administrators were keenly aware of how their positions within the system affected systemic reform, and that they usually held strong ideas concerning the best way to accomplish it. Policy implementation at the district and local school levels often means translating the aims of instructional reform from the district to principals and teachers.

In one of the few studies examining the role of district administrators in implementing systemic reform, Spillane (2002) used qualitative and quantitative methods to investigate the effects of state-mandated instructional policies on mathematics and science instruction in Michigan. His team interviewed 165 district administrators, teachers, parents, and curriculum specialists charged with developing instructional policies. A subsample of forty district administrator interviews was analyzed and the orientations of these administrators to teachers' professional development explored. Eighty-five percent of these district officials maintained a "quasi-behaviorist" perspective (Spillane, 2002) and assumed that highly prescribed professional development was necessary for teachers to implement mathematics and science reforms. However, their beliefs about professional development did not match the goals of the reforms. Instead, these administrators saw teacher learning from a behavioral perspective requiring adequate training from outside experts with external incentives for teacher participation.

In contrast, district change agents who supported a situated or cognitive perspective saw teachers as active agents in their learning who benefited from additional time for reflection. These district officials viewed teachers' day-to-day practice and student work as major components of the curriculum abetted by teachers' intrinsic motivation to work as experts and learners in their school communities. District officials who adopt a behaviorist perspective are not effective in supporting teachers' implementation of standards-based reform in contrast to district-level change agents guided by a situated or cognitive perspective who saw teachers as agents of their own learning in much the same way that standards-led mathematics and science reform views students' roles in learning requiring their active engagement.

Decision makers in any system are extremely important to effect-ive policy implementation. While organizational change can be initi-ated from anywhere in the system, those in positions of authority have command of the resources necessary to fuel reform and to influence the direction reforms will take. In this research district administrators were the highest level of decision makers whose perspectives we sought in conjunction with the Urban Systemic Initiative reforms. Their roles in the administration and facilitation of local school policy situated them uniquely between policy mandates and policy implementation. It was this unique status as both planners and implementers of reform that seemed to be a deciding factor in the final shape taken by reform in each site. District administrators must mediate the worlds of school leadership and policy implementation, often confronted with resolving the para-doxes of mandated school reform and site-based management. Similarly, principals referee the goals of school reform and implementation at the school. Together, district and school administrators have the power to influence policies that may determine the success of a reform initiative.

DISTRICT AND SCHOOL ADMINISTRATOR PARTICIPANTS

During the course of our work we interviewed a number of indi-viduals who hold a wide range of responsibilities for reform, as shown in Table 2.1. District administrators interviewed in our study included superintendents, as well as the mathematics, science, and technology specialists who were directly involved in implementing the systemic reform agenda. These specialists typically had twenty to thirty years of experience. Of this group, only one specialist had elementary teaching experience and the majority had taught fifteen or more years at the high school level. Directors were more diverse in their backgrounds, with many having elementary-level teaching experience; less than 20 per-cent had teaching experience in mathematics or science. Several of this group had no experience teaching in K–12 public schools.

These included university staff, research and evaluation staff, and staff at the district with control over the budget, all of whom played key roles in the implementation of the grant in several of the sites. Across the four sites, 47.6 percent of the district officials interviewed had held their current positions four to five years, placing them within the time frame of the USI reform efforts. Another 23.7 percent had held their positions six to twelve years, allowing them to compare current reform efforts with their district's approach to education before the introduction

Table 2.1.
District Administrator and Principal Demographics

	Chicago District (n = 8)		El Paso District (n = 10)		Memphis District (n = 8)		Miami-Dade District (n = 16)		Total District (N = 42)	
	n	%	n	%	n	%	n	%	N	%
Gender										
Male	2	25.0	4	40.0	2	25.0	5	31.3	13	30.9
Female	6	75.0	6	60.0	6	75.0	11	68.8	29	69.1
Race/Ethnicity										
Caucasian	4	50.0	4	40.0	2	25.0	8	50.0	18	42.9
African American	4	50.0	0	0.0	6	75.0	1	6.3	11	26.2
Latino	0	0.0	6	60.0	0	0.0	6	37.5	12	28.6
Asian/ Pacific Islander	0	0.0	0	0.0	0	0.0	1	6.3	1	2.3
Title										
Subject Specialist	3	37.5	0	0.0	3	37.5	2	12.5	8	19.0
Director	3	37.5	7	70.0	4	50.0	9	56.3	23	54.8
Super- intendent	2	25.0	3	30.0	1	12.5	5	31.3	11	26.2
Years in Position										
<4 years	1	12.5	2	20.0	2	25.0	2	12.5	7	16.6
4–5 years	5	62.5	6	60.0	1	12.5	8	50.0	20	47.6
6–8 years	0	0.0	2	20.0	2	25.0	3	18.8	7	16.6
9–12 years	0	0.0	0	0.0	1	12.5	2	12.5	3	7.1
Unknown	2	25.0	0	0.0	2	25.0	1	6.3	5	11.9

of the USI. Of those district administrators who felt most positively about the effects of educational reform, approximately half had been in administration less than five years. Although there is wide variation in experience, most directors averaged about six years in their positions. Most superintendents and associate superintendents had previously been teachers, assistant principals, principals, and district or regional administrators. Our interviewees included: (a) superintendents—all participants whose title included "superintendent," (b) curriculum specialists—mathematics, science, or other curricular and special projects facilitators, and (c) directors—administrators and directors of school system divisions and USI officials.

School principals have primary responsibility for facilitating the implementation of systemic reform at the school level. As such, a major focus of our work concerned how the forty-three principals we interviewed out of the forty-six schools participating in our research were involved in carrying forward a reform agenda. School principals in the study's sample had served in their current schools an average of six and a half years, with one principal having twenty years of continuous service in the same school. Thirty-nine percent of the participating principals were former elementary school teachers. They were an ethnically diverse group: 36 percent black, 21 percent Latino, and 43 percent white; none were Asian or Native American. Slightly more than half were females (51 percent).

Administrators at each level responded to the demands of implementing and sustaining reform in the context of their position's responsibilities, filtering their impressions through the lens of their experiences (Amatea, Behar-Horenstein, and Sherrard, 1996). Different perspectives held by our participants affected interactions between administrators and their implementation efforts, making each administrator's "perceived role" an important dimension for understanding systemic reform.

District Administrators' Involvement in Systemic Reform

When reform is given high priority, district policies are likely to sustain an environment conducive to systemic change. In general, however, those at the top with whom we spoke were not specific as to where most responsibility for reform rested. Was reform the major responsibility of the state, the district, or local schools? There was no consensus. However, district administrators were quite specific as to where their own responsibilities fell in the chain of implementation. Some saw the state in charge, primarily because of its role in high-stakes student assessment, and believed they were merely implementers of policy directives. They did not see themselves as change agents, but as go-betweens passing along directives from the state to the schools.

The majority of these administrators thought school principals were most responsible, believing the district's position was to oversee the work carried out by schools and engage in quality control. A handful understood reform implementation as teacher-driven and located in communities of learning in schools.

This finding is cause for concern—administrators at the top saw themselves as middlemen and generally did not see themselves as inhabiting the office where the buck stopped. How systemwide is systems reform when this response is the norm?

In carrying out systemwide reform, district administrators gave three areas relatively equal weight as primary vehicles for systemic change: *resource allocation* (65 percent), *professional development* (62.5 percent), and *curriculum requirements* (65 percent). Pressure on the district and schools for achieving high scores on standardized tests strongly effected implementation of the reform in each of the districts. Teacher certification and site-based management were also mentioned, but in one site—Chicago—almost exclusively. It is clear that the emphasis given to different strategies varies by city. For example, teacher certification is a much larger issue in Chicago than in other research sites, while resource allocation is most prominent in Miami-Dade, suggesting the difficulty this district faced in addressing the multiple needs of its highly diverse students.

In defining how reform outcomes related to district efforts in implementing systemic change, a superintendent in Chicago said it best when we asked him about the superintendent's role in the process of implementing reforms:

> And, of course, all decisions are made . . . whether they're budgetary, program, whatever . . . on the basis of performance—profit. And we measure . . . we measure profit by academic improvement. Not only measure [it] by test scores but also measure [it] by graduation rates and dropout rates and truancy rates and things of this nature . . . it's a management system that is designed to spend money in areas that promote . . . achievement. (CH08)[1]

Chicago, like all urban districts across the United States, is faced with accountability pressures from parents, students, and members of the community who will not tolerate student failure.

Resource Allocation.   Resource allocation includes funding for any reform-focused effort, from professional development to technological tools such as computers and graphing calculators. Some researchers (Cohen, Raudenbush, and Ball, 2000) maintain that in taking a view of teaching and learning as highly interactive (as we present in chapters 5 and 6), resources do not count for much unless they are actively used by teachers in framing tasks for students and unless students in turn understand and engage these tasks. According to Cohen, Raudenbush, and Ball:

> Ample school budgets will have little constructive effect on learning if they are not used to hire good teachers and enable them to work

effectively. Independent observers might report that such classrooms and schools had rich resources, but their potential to affect instruction would be unrealized. (2000, p. 12)

Resource allocation took many forms, but the most commonly discussed focus of district reform by those we interviewed was professional development for teachers. The responsibility for maintaining a teaching force current in educational methods and content was seen as a district-level responsibility and as one of the few ways that district administrations could directly affect student achievement outcomes. Resource allocation proved to be a more prominent concern for the larger school districts, Chicago and Miami-Dade. This notion is important because each USI site received 15 million dollars regardless of its size.

Several curriculum specialists emphasized the importance of ongoing support and resources in the classroom to assist teachers in using new mathematics and science curriculla and technology to support students' learning in accord with national, state, and local standards, recalling the interactive view of teaching and learning described by Cohen, Raudenbush, and Ball. In Chicago, for example, a curriculum specialist explained her role in reform in the following way:

> I think the best way to improve math and science in the district is to serve our teachers. The interaction that occurs between the teachers and students in the classroom is very valuable. So, it's that empowerment that I believe will definitely change the system. (CH07)

This specialist underscored the salience of classroom interaction between teachers and students and among students in small problem-focused working groups as the crucible for student learning. Many in Chicago's district offices saw their mission as empowering teachers to participate in the reform effort by providing materials critical for effective classroom instruction. Not all specialists saw reforms taking hold uniformly in all schools throughout the district, however. Top-down reform was not effective in institutionalizing change or empowering teachers as far as some were concerned. In describing the lack of well-established standards-based reform the specialist noted:

> Teachers don't have a real voice in professional development. They would do it because they've been told to do it and, at this point with the standards movement, teachers are like zombies, they're moving through it, and they're moving through it because they have to. (CH03)

This curriculum specialist saw teachers' compliance with USI reforms as less than wholehearted. In her opinion teachers put forward only minimal effort, burdened by their other responsibilities, which did not disappear when USI and standards-based reforms were put in place. Another curriculum specialist in Chicago agreed, saying:

> We have implemented standards-based examinations at the high school level. However, it fell short of what needed to be done. Just developing curriculum documents which say that these are the standards of the Chicago Public Schools and not giving the teachers the tools with which to implement them falls short. (CH01)

While progress had been made, teachers did not receive enough support in the form of additional professional development and resources to allow them to implement successfully the new standards in their classrooms. Administrators in Chicago were especially concerned that professional development was not solving the problem of reform implementation, while curriculum specialists thought teachers did not receive enough consistent support to carry out a constructivist approach in their classrooms. An area supervisor in Chicago put it this way: "[It is all about] utilizing standards and a framework that is consistent with insuring the substance of what is taught grade by grade and level by level is consistent with what those standards say ought to be happening." Those closest to the teachers—the curricular specialists and regional office coordinators who most often interact with the schools—focused on the lack of sustained assistance and resources at the classroom level.

Resource allocation was discussed most among district administrators in Miami-Dade. According to one Miami-Dade administrator, "Policies in this district, almost in every sense, encourage student achievement. The policies are not the barriers. The barriers are [the lack of] resources" (MI14). In Miami-Dade resources needed to support professional developed were most often discussed. Along with the reallocation of Title 1 and Eisenhower funds, the USI grant enabled Miami-Dade to support professional development opportunities throughout the district. Administrators believed that this leveraging of funds made the most difference.

Professional Development. Virtually all associate superintendents in charge of curriculum and instruction saw professional development as most powerful when it included both teachers and administrators in either on-campus mentoring sessions or off-campus workshops. The link between professional development and resource allocation was mentioned frequently. Professional development that included the provision of content knowledge and pedagogy, as well as the materials needed to carry it off, was the key to accomplishing systemic change. A professional development specialist in Chicago described his work with teachers as critical for connecting teachers to others, to the standards, and to the curriculum, making sure that students are learning what teachers are teaching and that what they are teaching is tightly connected to rubrics and standards:

> We do in-services for the teachers. We bring them to workshops at De Paul where they meet other teachers from other schools and share strategies, come back and implement them, connect with other teachers at their grade level, or in their department . . . We work to support teachers in order to ensure that what they're teaching is what students are learning. And to make sure that what they're focusing on is what's important, and what's articulated . . . in the state goals and academic standards, [the] framework statements. And so we try and simplify it and make it happen. (CH100)

However, according to a curriculum specialist in Chicago:

> All we do is create frustrated teachers by not giving them an adequate budget to buy the materials that they need to have a good hands-on science program. We teach them about the efficacy of hands-on and about integrating science with math and then we send them back to their school and they have nothing to work with. I think this is unconscionable. . . . It [CSI's math, science, and technology improvement program] provided opportunities for teachers to get the professional development but it provided no funds for the schools to buy instructional materials for students. It's like teaching students to play the piano on a paper keyboard. (CH01)

A senior science specialist in Chicago who had been working with the CSI since its inception repeated the argument that the only effective strategy was to build capacity at the school level by creating teams of committed teachers:

> And after working with schools one on one for two years, it became apparent that the only way we were going to sustain what we did was to come up with a model where schools could professionally develop themselves. In other words, this whole notion about long term professional development is a very critical issue because it was apparent to us that once teachers in the building have been professionally developed we had

no notion of what they did with that. Did they share it with other teachers? Did they implement it within their classrooms? So, we took another approach to building some capacity within the schools; we developed this professional development team the third year of CSI, and we worked with the design team leaders and a team of teachers from the school. And we had professional development strategies on an ongoing basis that took more of a team approach with the schools, as opposed to an individual approach for the individual teachers. (CH01)

Miami-Dade took a very different approach. During the first two years of their USI reforms they implemented what one interviewee referred to as "massive training in summer sessions" held in a central location. By the third year of the USI reform, however,

We brought together some math and some science teachers from each school to work out a feeder pattern plan that was implemented as our method of professional development during the year. (MI11)

As Miami-Dade's USI recruited additional schools each year for participation in the reforms, they implemented a feeder-pattern approach. The plan pulled teachers from mathematics and science from different schools to work together to support schools and teachers linked together in the same region of the district. By the end of the USI grant principals selected mathematics and science teachers from a feeder pattern who met with USI district staff to establish a professional development agenda addressing the consensus needs of the feeder pattern. District staff felt that they could support the needs of twenty-five feeder patterns more effectively than they could support 300 independent schools.

The issues driving policy changes in both Chicago and Miami-Dade were similar nonetheless: teachers needed more time to reflect upon and use the strategies they learned in professional development and the limited number of district staff needed a feasible mechanism for providing support for large numbers of teachers and schools. The initial one-shot summer sessions failed to provide teachers the necessary experience with new approaches to teaching and learning that systemic reform entails, and the limited number of district staff could not address the specific needs of the districts' many schools. As we will see in subsequent chapters, Miami-Dade teachers, more than most teachers in the three other sites, struggled with implementing fundamental changes in their teaching practices in line with the reforms.

Curriculum Change.   The importance of combining resource allocation with professional development and challenging curriculum is an idea that was widespread in district offices. Increasing and maintaining challenging curriculum requirements in mathematics and science for all students has become a prevailing concern. With standardized testing and reform efforts pushing schools to be more productive (as reflected primarily in student achievement outcomes), many districts are raising the bar to insure that students gain more sophisticated skills and knowledge in mathematics and science (in addition to language arts/reading). For example, to raise academic standards, algebra was institutionalized as a course requirement for graduation in Miami-Dade.

The creation of challenging curriculla aligned with assessment systems was most apparent in Texas and to a lesser extent in Florida. In Miami-Dade we conducted a focus group discussion with curriculum specialists who had attended an assessment workshop in Texas and remarked, "they have a system in place which is a phenomenal statewide system called Accountability in Student Achievement." Districts and schools in Texas were expected to show a year's cognitive gains for students at all grade levels every year and is, with the passage of No Child Left Behind (US DOE, 2001), now a national requirement. This accountability system aligns with the USI goal of implementing systemwide changes to combine the efforts and goals of district policy into a unified whole. Aligning assessment with curriculum is an effective tool for infusing instructional quality into classrooms and promoting student achievement. An associate superintendent of instruction in Memphis followed the alignment of policies from professional development through the curriculum, supported by resources, in the following way:

> Teachers need an opportunity over time to come for training . . . during the day when they are on the payroll, where the expectation for their performance is always higher than on their own time. There is time in between that and the next session where they can apply what they learn. They have materials that support them. They have follow-up for trainers that are here that are in their classroom. They come back having used the ideas, used the material for the next session. (ME06)

The support of the district through provision of long-term professional development and the implementation of standards-based curriculum both allow teachers to incorporate standards-based instructional strat-egies into their daily classroom practices in line with an interactive view of teaching and learning. A science facilitator in Memphis envisioned reform as the responsibility of those in the schools and believed the role of district administrators was to provide information on national trends and state initiatives. This director saw herself as a provider of knowledge rather than as an active participant in the process:

> As science facilitator, my responsibilities are to ensure, as much as possible, that schools are able to facilitate changes, or bring about changes in the science education program. So academically, it's keeping the district current on national trends in science education, such as the education science standards, performance based activities and standards based curricula, things that emerge over time. In order to keep the district well informed about what's going on in science with assessment and so forth. And also act as consultant to schools as the local site when they request it. (ME02)

An assistant superintendent in Miami-Dade agreed that meaningful reform takes place at the school level, but enlarged the district's role to include teacher preparation and professional development. He believed that the district was responsible for maintaining teaching quality through established relationships with teacher preparation colleges and novice teachers in addition to the district's ongoing support of established teachers. He asserted that the district's responsibility was to the teachers. He did not see principals and other school administrators as part of the equation:

> Well, you have to look at it from three levels of teacher preparation. One, obviously, is pre-service. If you look at pre-service then you need to have good working relationships with your local colleges and universities. So that you can work cooperatively to help the prospective teacher be well prepared when they enter. Then you have the period of time known as "induction" when you have the new teachers who have come into the system, who need more generic support for teaching in general and maybe a little less in terms of content. And then you have the in-service for those teachers who have been here awhile and who need to be provided opportunities to stay current and improve their skills. So, I see it in those three areas. The district has a responsibility in each of those areas. (MI11)

While district administrators such as this one in Miami-Dade defined roles for themselves in supporting teachers' acquisition of skills and knowledge to support reform, they did not often envision a network of support that included principals and other building-level staff working together to accomplish changes in instructional practice that supported subsequent student achievement.

Assessment

High-stakes tests administered annually such as Florida's state-mandated Florida Comprehensive Achievement Test (FCAT) are used with increasing frequency by states and districts to evaluate schools and those who work in them and make funding decisions. Alternative and supplemental achievement measures are also common; nonetheless, a continued emphasis on these "high-stakes" tests within schools prevails due to the importance placed on student scores at the state and national levels.

One director in Chicago took this emphasis even further, adding that:

> Our desire is for 80 percent of the children in this district to score well. Now when we achieve that, what population are we talking about? We're talking about the minority population. When at least 80 percent of the children in this district, African American and Hispanic, are scoring at or above national marks, then we will be significantly impacting underserved children. (CH07)

For reform to be truly successful in Chicago, not only 80 percent of students, but also 80 percent of minority students must score at or above national averages. This is a key point because, to realize systemic reform, the achievement gap between underserved and other students must be narrowed. Although this director placed caveats on the interpretation of achievement percentages, she believed they presented an accurate portrayal of student achievement in Chicago's public schools.

Unfortunately, standardized tests place pressure on teachers to focus on a narrow set of topics that they understand will be a major focus of the high-stakes test. A director in El Paso sees this as the case in her district:

> I think the districts have been very good in implementing . . . in prioritizing mathematics, beginning now, I would say, and science, but primarily mathematics, because what gets tested gets done. (EP02)

She saw gaps in the implementation of reform as a result of emphasizing mathematics over science in carrying out the changes and restructuring associated with the USI. This director approved of the district's overall plan of implementation, but emphasized that "what gets tested gets done." This view was commonly held by district-level admin-istrators who, while recognizing the usefulness of student test perform-ance in deploying resources, including opportunities for professional development, also recognized that complete reliance on test score data to judge resource needs might undermine the success of the reform, especially if teachers "taught to the test" through drill and practice.

The pressures of standardized testing on the schools are twofold. While principals were asked to implement reforms that rely more on "hands-on" and "exploratory" learning, those skills may not be reflected in standardized assessments. An associate superintendent for instruction in Miami-Dade believed that because standardized tests were used as a measure of both student performance and school performance the test was emphasized at the cost of other important educational experiences:

> I think the bigger danger is the emphasis on standardized testing. Because there are many things that tell me that a student has an under-standing of a subject besides their performance on a standardized test. A standardized test can never, in my estimation, be the sole measure of a student's learning. There are just too many other things. And the risk, as I see it, is that when the standardized test drives an accountability system that labels schools and kids that the pressure becomes such, on teachers in schools, and administrators, that there is an over-emphasis on "the test," as opposed to a rich instructional experience. (MI14)

The idea that a student can have an understanding of subject matter that standardized tests will not assess led district administrators to sup-port alternative means of assessment. In Chicago a director suggested forms these alternative assessments might take:

> It could be a paper pencil assessment. It could be an interview. It could be a group discussion in class, where the teacher is standing there and pro-cessing, very quickly, all of the information that you are gaining from the general conversation that you're having with the students, okay. And of course, it could be artifacts such as a portfolio. (CH07)

Overall, the decision makers we interviewed thought their policies were well implemented at the local school level and saw their primary task to be supporting national and state policies in order to increase student achievement. The intricate connections among resource allocation, professional development, and curriculum requirements were the primary district vehicle for enacting curricular change at the classroom level.

Administrators at different levels in district offices agree on tactics to increase reform success but differ in what they see as indicators of success. Superintendent and director-level administrators spend little time with school-level staff and offer sanguine accounts of reform implementation through professional development activities. But curriculum specialists, who are often called on to interpret standards and forge plans for shaping practice at the school level, see a great need for increased teacher support. We next follow the path of reform implementation into the schools. In each school the principal embodies the administrative authority to guide a vision of teaching and learning. Principals are the decision makers held most accountable for their teachers' and students' success.

School Principals and Reform

We interviewed principals in each of our participating schools. A primary focus was the principal's assessment of how best to carry forward mathematics and science reform. The topics considered most important by principals in undertaking successful implementation of reform in their schools included: (a) professional development; (b) school demographic factors such as student ethnicity, language use, and socioeconomic status; and, most importantly, (c) school vision, attitudes, and guiding principles supporting a culture of reform.

A supportive school culture facilitated by principals who saw themselves as both instructional leaders and members of a larger school community was an indicator of student mathematics achievement gains during reform implementation, a topic we explore in greater depth in a subsequent chapter. In addition, many principals mentioned the importance of support from the surrounding community, including teachers, students, district-level administrators, parents, and community members, an issue we return to in the next chapter.

We also asked each principal to help us understand his or her role as building administrators and heads of site-based management within their districts. Close to 50 percent viewed instructional leadership within their schools as paramount, with smaller numbers seeing professional development (31.8 percent), community collaboration (27.3 percent), and facilitation (22.7 percent) as critical. Rather than simply focusing on a bureaucratic approach to the management of their schools, principals viewed themselves as both human relations managers and curriculum leaders who mediate multiple levels of stakeholder interactions, maintain high academic standards for all students, and simultaneously provide equal attention to resource allocation, reform initiatives, teacher concerns, and public relations.

We assume that principals can and do influence policies emanating from the state or district. Principals therefore shape a school culture that either supports or undermines reform. We believe that vibrant school cultures are associated with positive outcomes for students and that student-centered principals are an important element of this mix. Our perspective draws from Danzig's (1999) view that student-centered leadership suggests an ability to frame educational issues from multiple perspectives and to minimize the damages stemming from simplistic views, misguided views, self-defeating views of children, or in maintaining the status quo.

> Learner-centered leadership involves a balance between the professional norms and personal dispositions of educators, with the larger good as defined by a learning community. Without this focus on learning, there is considerable risk that the daily press of management tasks and a crisis mentality will override the school leader's attention. This enlarged role of leadership implies a movement away from bureaucratic models to a post-industrial model of schooling with the goal of educating all youngsters well. Two challenges that exist will be to reorient the principalship from management to leadership and to re-focus the principalship from administration and policy towards teaching and learning. (Danzig and Wright, 2002, p. 6)

During the first year of our three-year project, in addition to interviewing principals we also asked faculty members to complete short surveys investigating their perceptions of factors contributing to the quality of the school culture. Other school-level data included student achievement in mathematics as measured by student performance on norm-referenced tests such as the Stanford-9 and on high-stakes tests such as the Texas Assessment of Academic Skills (TAAS).

As the last step in policy making and the first in policy implementation at the school level, principals are directly responsible for the implementation of reform in their schools and uniquely vulnerable to accountability, including standardized testing. Earlier in this chapter we mentioned that central office staff pointed to the principal as a driver behind successful implementation of reform in schools. Most principals agreed that their vision and goals influence student achievement and motivate teachers to incorporate standards-based practices into their classroom practices.

BARRIERS TO REFORM

We also spoke with principals about factors that acted as barriers to reform. In a previous analysis of our principal interview data (Kersaint, Borman, and Boydston, 2001), we determined that principals in each of our four participating USI sites were preoccupied with three major issues or contradictions that beset them in their day-to-day work in implementing USI reforms. First, principals perceived a mismatch between the goals of the reforms and the objectives of state and school district accountability practices. Although NSF reform emphasizes hands-on problem-solving approaches to mathematics and science, high-stakes testing can promote rote learning and regurgitation of facts because many teachers do not trust hands-on, problem-solving strategies to get the job done. In McNeil's (2000) terms, testing and standardization also foster defensive teaching of "fragmented and narrow information on the test which comes to substitute for a substantive curriculum in the schools of poor and minority youth" (p. xxvi).

Second, principals confront the practical concern of finding substitutes for classroom teachers who attended professional development sessions off campus. While this may seem to be a trivial issue, for principals faced with managing instructional resources, facilitating the professional growth of their teaching staff, and enhancing student learning, it presents a major challenge in implementing reform. Professional development is the cornerstone of most large-scale instructional reform, and USI reforms are no exception.

Finally, principals were concerned about the pressure to take on new entrepreneurial responsibilities in funding reform. Several talked about using other funds to augment NSF resources in purchasing new computers and other technology. All were burdened by pressure to create working capital to supplement allocations from the school district and other sources.

Assets in Reform

While principals occasionally saw themselves burdened by difficulties in implementing reform in mathematics and science in their schools, they more often saw themselves as active agents taking on a number of roles to support reform. In Table 2.2 we see that principals' perspectives on their day-to-day work illustrate the differences we recorded across sites. In Chicago professional development and facilitation were most prominent; however, relatively few principals in El Paso mentioned these aspects. In El Paso principals focused on encouraging

Table 2.2.
Principals' Perceptions of Their Role in Reform

Role	Chicago (n = 8) %	El Paso (n = 8) %	Memphis (n = 12) %	Miami-Dade (n = 16) %	Total (N = 44)%
Instructional Leadership	25.0	62.5	66.7	37.5	47.7
Professional Development	50.0	50.0	25.0	18.7	31.8
Community Collaboration	37.5	62.5	16.7	12.5	27.3
Facilitation	50.0	0.0	16.7	25.0	22.7
Policy Implementation	12.5	12.5	33.3	12.5	18.2
Diverse Roles	12.5	12.5	16.7	18.7	15.9

community collaboration instead. Generally, principals put effort into those aspects of their roles they see having the biggest payoff.

The principals' focus on instructional leadership and professional development overlaps with district administrations' understanding of professional development and curriculum as different manifestations of the same policy base. Principals do not have as much power to set curriculum requirements as district administrators, but they are responsible for acting as instructional leaders and encouraging teachers to implement the higher standards and curricula enshrined in district policies. Also, as district officials channel resources into professional development, principals react to this by supporting professional development activities to increase teacher effectiveness, but also by securing substitute teachers who are able to teach effectively.

Facilitation, policy implementation, and the mediation of diverse roles and responsibilities are secondary aspects of principals' day-to-day work, usually accompanying instructional leadership and professional development. More emphasis on policy implementation was noted in Memphis and could be explained by the extremely active reform administration in Memphis City Schools under Superintendent Gerry House during the mid- to late 1990s. Each school in the Memphis district was required to adopt a reform model and work toward reform goals on a district-wide schedule.

INFLUENCE OF ASSESSMENTS

Student assessment—both how assessments were made and what the consequences were—affected virtually all school principals in our study. This is in contrast to their district-level counterparts, only 20 percent of whom stated that standardized testing was very important. Many principals are under considerable pressure to produce ever-greater percentages of students performing at high levels in mathematics, science, and reading/language arts. All of the principals in Chicago and El Paso and all but one in Miami-Dade discussed the influence of high-stakes standardized testing on instruction and accountability programs in their schools. Eighty-three percent of the principals in Memphis did not emphasize high-stakes assessment in their interviews. Memphis district administrators similarly had a low rating for the influence of standardized testing, at 12 percent. This differential in the emphasis placed on standardized assessments are likely due to the accountability policies in each of these states.

The primary means of judging the success of reform implementation is student achievement as an indicator of program efficacy. To measure student achievement, interviews, group discussions, and portfolios were all mentioned as assessment tools that could be used by teachers, but some principals (and their district administrator counterparts) were skeptical about relying on portfolio assessments because such practices ran counter to their state's methods of assessing student achievement.

Many principals reported that a successful way to improve students' scores on high-stakes assessments is to invest in programs for low-achieving students. Sixty-one percent of all school principals reported the use of special academic programs to increase student achievement focused on the lowest achieving students. These programs relied almost exclusively on academic tutoring, with only 7 percent addressing economic concerns and 4.5 percent addressing social concerns. Economic programs such as free and reduced-price lunches were often referred to as well, even though these are not specifically aimed at improving student achievement.

CONCLUSION

In sum, district and school administrators alike saw resources, including support for both professional development and curricular requirements, as critical to enhanced student achievement outcomes. Resource allocation was particularly challenging in large school districts because they were using limited funds to address the needs of a large number of schools and teachers. This suggests that although 15 million dollars is a substantial amount of money, its influence is limited by the size of the district. Additionally, leadership at the district and school levels is at the heart of systemic change in schools and classrooms. This leadership is manifest in their ability to facilitate professional development activities as central to reform. Many principals cited the promotion and coordination of professional development as part of their responsibilities as instructional leaders. Superintendents and others at the district level as well as principals are under enormous pressure from states, federal programs, community constituencies, and others to be accountable and productive. Accountability in the form of high-stakes assessments guided district and school administrators' decisions about allocation of resources and the reform effort at the school level. Close to 100 percent of all interviewed principals stated that standardized tests guide instruction and accountability in their schools. Results posted by students on high-stakes assessments such as FCAT in Florida, TAAS in Texas, Tennessee Comprehensive Assessment Program (TCAP), and Illinois Goals Assessment Program (IGAP) are used by administrators at the school and district levels to allocate resources aimed at enhancing instruction and improving student achievement. For example, district administrators interpret results on high-stakes exams to inform district policy on curriculum requirements such as whether Algebra I is offered in the eighth grade, while school principals attempt to customize programs to support the individual needs of a given school's population. Because of this, individuals occupying these positions are expected to juggle diverse and often conflicting sets of responsibilities, making their jobs especially demanding. The next chapter describes how principals, district administrators, and community stakeholders work to build relationships among schools and communities.

3

Building Relationships to Sustain Reform

We really are not supposed to be a social service community, we are a science community. But we realize that sometimes obstacles are standing in the way of achieving in math and science and other subjects. So you have to address those to be able to effect change in the academic areas.

—Miami-Dade research participants

This statement from two Miami-Dade community stakeholders demonstrates the fact that when stakeholders become involved in educating students, their involvement can take any number of forms, from academic to social or something in between. A dramatic increase in awareness concerning the relationship among home, school, and community became evident in education reforms of the 1960s and 1970s (Cairney, 2000). These reforms pointed to the importance of parental and community involvement in insuring successful outcomes for students. A host of initiatives, including comprehensive school reform and systemic reform, in the 1980s and 1990s included community involvement as a key component in implementing school change.

Research has indicated that stakeholder involvement creates benefits for students by encouraging cognitive readiness (Reynolds, 1991), the expectation or belief that they can succeed (Okagaki and Frensch,

47

1998), and the motivation to do well (Reynolds, 1991). According to Saphier and King (1985), school improvement emerges from the strengthening of teachers' skills, the systematic renovation of curriculum, the improvement of the organization, and the involvement of parents and citizens in responsible school-community partnerships.

Stakeholder relationships are critical in supporting educational reform in general and mathematics and science reform in particular (NSF, 2000). The significance of stakeholder involvement is underscored by research that demonstrates connections between school-based partnerships and outcomes for students, parents, schools, and school districts. The NSF Six-Driver Model, described in chapter 1, includes Driver 4, "Mobilization of stakeholders," emphasizing the importance given by the NSF to collaboration and cooperation within districts and between schools and the larger community.

In this chapter we discuss how community and parent stakeholders participate in schools involved in the USI reform in an effort to address a research question that guided the policy study: How are resources mobilized among various constituencies such as schools, universities, business and industry, and political agencies?

To uncover the extent to which community and parent stakeholders are involved in the USI—particularly with the schools participating in our research—we look at findings from a variety of data sources gathered during the course of our project. The results are discussed based on the large range of stakeholders from the district administrator level down to the community and parent level. First, we present findings that reveal how community and parent stakeholders have been mobilized within the USI. That is, we discuss the activities of stakeholders and the individuals or groups responsible for supporting student learning in the schools. Second, we describe the relationship of stakeholder involvement to the reform initiative. Finally, we discuss the perceived relationship between stakeholder involvement and the intended student outcomes.

ENGAGING STAKEHOLDERS

Systemic reform reaches beyond the classroom and the school to engage the participation and cooperation of a wide variety of individuals and organizations. Although rates and types of participation varied within and across school districts, each USI site witnessed the involvement of a wide range of stakeholders. Stakeholders, in the context of this study, are individuals and organizations with an investment in education and who are participating in the systemic reform effort. Stakeholders include district personnel, principals, teachers and other school staff, students, parents and guardians, and individuals from businesses, faith-based organizations, and government and community agencies.

At all levels, and in all USI sites, stakeholders spoke of the value of building relationships among schools, parents, and other partners. Although usually focused on their own school, neighborhood, or district, stakeholders from diverse backgrounds discussed their involvement, goals for involvement, and impact on student achievement. This section examines the relationships that developed during the course of the USI by discussing the results from the analysis of district and school policy documents, interviews with district and school administrators, and, most importantly, from community stakeholders themselves.

District-Level Mobilization of Stakeholders

By 1998 three of the sites in our study had finished their fourth year of funding: Chicago Systemic Initiative, El Paso Collaborative for Academic Excellence, and Miami-Dade Urban Systemic Initiative. Because the Memphis Urban Systemic Initiative was funded as part of NSF's second cohort of district sites (1995), this site was one year behind the other three school districts in implementation of the reform. Each site involved in our study used very different strategies for maximizing broad-based community support. The Chicago USI not only partnered with a number of businesses and community organizations but also organized a collaborative between the Chicago-area museums and universities so that together they could improve services to students and teachers. In El Paso progress toward building relationships with the community had begun in 1998 (El Paso's fourth year of funding), with the El Paso USI collaboration in the midst of establishing a new business/community initiative. The Miami-Dade USI focused on increasing communication with the community by distributing more than 100,000 brochures and pamphlets as part of its broad-based outreach program (Miami-Dade County Public Schools, 1998), while the Memphis USI encouraged stakeholders to visit schools to observe mathematics and science activities.

For their efforts, an increase in the level of support from community stakeholders was reported from all of the sites except El Paso, which did not report any increases in stakeholder involvement since it had just begun to establish stakeholder relations. The Chicago USI reported an increase in collaboration among the stakeholders it assembled.

Miami-Dade USI experienced an increase of support in response to its mass distribution of brochures and pamphlets to stakeholder groups throughout the district. According to Miami-Dade's 1998 annual report, new community-based mathematics and science projects were initiated in response to and in direct support of the USI. The Memphis USI experienced a substantial increase in financial support from community organizations for the reform.

The four USI sites shared the common goals of improving the achievement of all students and having relevant stakeholders understand and accept systemic change as a strategy to improve mathematics and science education. In order to increase the achievement of all students, one policy change that resulted from the Memphis USI was to eliminate low-level science and mathematics courses. Most Memphis stakeholders supported this policy and saw the potential for all students to complete a rigorous regimen of science and mathematics (Memphis City Schools, 1998).

The Chicago USI, guided by the belief that "all children can and must learn to their fullest potential," was compelled to engage actively and continually all stakeholders, teachers, administrators, parents, and the community in this endeavor (Chicago Public Schools, 1998). Each partner was responsible for assisting with the implementation of at least one, if not all, of the USI goals and/or benchmarks. Parents and community groups attended ongoing workshops presented by parent and community partners to recognize standards-based instruction. Workshops included Parents Assisting Local Schools and Parents as Teachers First. According to the Chicago USI Annual Report (1998), all stakeholders, teachers, administrators, parents, and the community were actively and continuously engaged in CSI.

The Miami-Dade USI emphasized the importance of community-based organizations providing curriculum materials, classroom programs, and field trips to support the USI. Representatives from the Miami-Dade USI reported, "Parents and community members are embracing mathematics and science through a variety of engaging activities" (Miami-Dade County Public Schools, 1998, p. 1). Through the Principals' Institutes, teacher leader conferences, and feeder pattern workshops, stakeholders learned the importance of making connections across all elements of the system's curriculum, instruction, assessment, policy, school improvement, and community involvement.

Across all four sites, 24 percent of district-level administrators discussed the importance of involving university partners in conducting in-service training and engaging pre-service teachers in the district's efforts. A Memphis district administrator's perception of university partners is representative of district administrators from all four sites:

> We work very closely with the colleges and universities in pre-service education. One of the things we want to do is to be sure that the course work that is offered at the university is consistent with the reform we have established in the district. And secondly, we want to assure that those prospective teachers have additional opportunities early on to actually work with the kids. (ME05)

Additionally, this administrator acknowledged the benefits of collaborating with university partners in training the district's teachers.

Another community involvement theme that emerged was the emphasis on preparing parents to make informed decisions about their children's education as well as involving them in the reform. Thirty-eight percent of Chicago district administrators, three of the interviewed Miami-Dade administrators, and two El Paso administrators perceived that "training" parents through outreach programs increased their ability to be involved in their children's education and school. The assumption is that "[parents would] be in a better position to make informed decisions about the relevance of our school if they are a part of professional development" (CH03). District administrators perceived that involving parents in professional development to understand the reform goals and assist their children with mathematics, science, and technology was far more important for students' subsequent achievement than attending Parent Teacher Association (PTA) meetings.

A third theme identified was the importance of preparing students through community involvement to succeed in post-secondary activities in a career, at a college or university, or in their current mathematics and science courses. District administrators believed that by strengthening technological, mathematical, and scientific skills, students would succeed in post-graduation employment. District administrators in Miami-Dade argued that preparing students for employment is a concern for not only the school but the community as well:

> When you look at some of the industries that are having a great deal of difficulty meeting a need for the area because of all the telecommunications and so forth, all of them prefer a certain amount of mathematics and science background, entry skills, computer skills, that type of thing. I think it's the concern of the whole community to turn out a student that is more academically advanced. (MI06)

This administrator echoed sentiments expressed across all four sites regarding the importance of preparing students for employment. In addition, most endorsed community projects as preparation for skilled performance on high-stakes tests as critically important:

> We have algebra camp in the summer for all kids who are going to enroll in algebra. And that spurned the geometry camp that we have now. We have ACT study sessions for kids. We have in the summer something we call STAT Camp, Science Technology Algebra Teams for seventh and eighth graders. Each one of those was an outgrowth of USI, and they've grown. (ME06)

Many district administrators believe that participation in summer school study sessions boosts student achievement on critically important high-stakes tests. Virtually all administrators shared the belief that stakeholders can be important educators whether the community is seen as universities, businesses, community groups, or parents.

SCHOOL-LEVEL REPORTS OF INCREASING STAKEHOLDER INVOLVEMENT

Each year a school improvement team, composed of the faculty and staff of a school, constructs a school improvement plan, or campus action plan, as it is called in El Paso. In the improvement plan policies and goals for the school year are described in addition to other topics, including mobilization of stakeholders. In this section we present the findings about how stakeholders are mobilized according to the policies and goals described in 1999–2000 school year School Improvement Plans (SIPs) from all participating schools in each of the four sites.

There were five types of partners that were identified in SIPs from the four school districts: community organizations, universities, businesses, churches, and parents. Chicago schools participating in our study had the greatest variety of stakeholder involvement, with each SIP mentioning at least two different types of stakeholders from the five categories. However, community organizations were mentioned most often, with seven of the ten schools identifying community organizations as school partners. Memphis schools also identified stakeholders in each category, but focused on the involvement of businesses rather than community organizations. While SIPs in El Paso and Miami-Dade emphasized parental involvement, SIPs in Miami-Dade had the lowest number of references regarding stakeholder involvement. However, unlike in

El Paso, parent partners in Miami-Dade SIPs were identified as "directly involved as volunteers, tutors, and mentors" (Miami-Dade County Public Schools, 2000). Nine of the ten El Paso SIPs had goals of "increasing parent participation," and "organiz[ing] and implement[ing] a successful PTO organization" (El Paso Independent School District, 2000). This suggests that the schools do not have parents participating in their schools and that structures may not be in place to enable parents to participate. It appears that the El Paso schools participating in our study were in the process of developing ways for parents and community members to become more active in their schools at the time of the 1999–2000 SIP.

In our Memphis schools participation of business stakeholders involved "assisting when needed through donations of incentives, participation in activities, and tutoring" (Memphis City Schools, 2000). Church organizations were also frequently mentioned as stakeholders. According to one school's SIP, its church partner "works with our staff and students on a regular basis to enhance the educational experiences for each student" (ME240). Two schools in Chicago also identified church organizations as partners, while none of our schools in El Paso or Miami-Dade mentioned church organizations.

Principal Views of Stakeholder Involvement

We interviewed forty-three schools' principals during the 1999–2000 school year. During the interview, principals were asked to describe how the participation of university partnerships, business and community organizations and members, and parents, among others, has impacted the school. In Chicago two-thirds of the principals mentioned community organizations as school partners, which mirrored what was reported in the SIPs from these principals' schools. These community partners included hospitals, museums, and counseling groups, as well as other groups broadly referred to as "community organizations." Community organizations are involved in Chicago schools in several ways. Museums invite teachers to evening events so they can learn about the programs that are available. Hospitals provide instruction to students, while other community organizations provide speakers for the school or counseling to students. One Chicago principal summed up, "You can't have enough community [involvement] because I think that's a key to improving a school" (CH210). In Chicago businesses and universities were also frequently mentioned as stakeholders in the schools. Businesses tended to provide nonacademic resources, such as refreshments for ceremonies and "give-away" prizes like t-shirts. However, one principal noted, "Most of our partnerships have to do with businesses in our vocational aspect of our school" and continued by stating, "we need to start pursuing partnerships in the areas of math and science" (CH310). University involvement in the participating Chicago schools tended to involve offering graduate and undergraduate students as tutors and mentors. Only two of the Chicago principals in our sample identified parents and churches as stakeholders.

In El Paso schools half of the principals referred to business and community partners during our interviews. Business partners included corporate businesses such as a major soft drink corporation or restaurant chain. Others were local business people, who, according to one principal, "come into the school on instructional issues" (EP310). Similarly, community organizations such as Lions and Kiwanis clubs were partners in education. Only two principals mentioned universities and colleges, one of whom stated, "UTEP [University of Texas at El Paso] is very important" (EP210). The other principal mentioned a community college's involvement: "They come out and judge our science fair and they do different things" (EP130). We found the omission by the other principals in El Paso to mention UTEP as a stakeholder in their schools to be interesting because UTEP is one of the main organizations that constitute the USI El Paso Collaborative. This could suggest that UTEP's ubiquitous role in the USI effort excluded them from being considered an entity separate from the USI. Additionally, El Paso principals in our study did not mention church organizations or parents as partners. However, parents were identified as stakeholders in nine of the ten SIPs from these El Paso schools.

In Memphis City Schools a third of the principals in our study identified business partners. A principal of an elementary school stated, "We have adopters from the business community that have helped us to fund lots of incentive programs for children in math, science, and reading" (ME130). Mainly, business partners in Memphis schools participate by giving funding for technology or providing mentors to students. One high school principal explained, "Four kids are interning over there [at a business] and their future looks really bright" (ME310). Three principals identified local universities or colleges as partners with their schools. Their involvement in the schools consists mainly of providing tutors. Likewise, one principal identified two local churches as partners who provide tutoring to students. None of the Memphis principals mentioned parents as partners.

Slightly more than half of the Miami-Dade school principals in our study identified university partners that assisted their schools. Participation from the university typically involves students taking coursework in mathematics, science, or technology during the summer and attending programs such as Math All-Star, which focused on engineering concepts. One principal stated,

> [The university] is involved in our Partners for an Academic Community and I think that this has a very large impact science- and math-wise on our kids because they are taught by professors at the university in math and science and then they come back to [school] for the remainder of the day. (MI330)

While university partnerships are strong in some schools, in others they were just beginning to grow. As another principal pointed out, "our relationship with [the universities] is beginning to strengthen. There's more interaction. It's beginning to develop" (MI320). Almost half the Miami-Dade principals in our study identified business organizations as partners. Business partners provide school-to-work experiences, technology assistance, and tutorial programs. One elementary principal identified a bank as a partner, stating, "[the bank] provides tutorial programs for students below grade level in mathematics and reading as well as financial assistance [to the school]" (MI110). Twenty-nine percent of our participating Miami-Dade principals also noted community organizations as important stakeholders. Like business partners, community organizations assisted by providing tutors and offering after-school learning opportunities for students. Only one high school principal mentioned the school's alumni association. This principal indicated that the alumni association was the "heartbeat" of the school, and that "very few high schools have it" (MI310). Indeed, none of the other forty-two school principals mentioned an alumni association as a partner for their school.

In sum, all four USI sites mobilized resources outside of the school. Chicago schools mainly sought partnerships with community and business organizations. In El Paso only a few principals identified school partners, and these were generally from community organizations and businesses. The apparent lack of partnerships between the school and the surrounding community coincides with district administrators from the El Paso USI, who stated that they began encouraging school-community partnerships by forming committees in 1998. In Memphis City Schools principals for the most part identified businesses and universities as school partners. Miami-Dade principals identified their school partners as the local universities, community organizations, and businesses.

COMMUNITY STAKEHOLDERS' VIEW OF USI IMPACT ON THEIR INVOLVEMENT

According to 70 percent of the community and parent stakeholders interviewed across the four sites, the USI has had little to no impact on their involvement. Involvement was determined by whether or not stakeholders participated in a USI program or received funding from the USI to support their programs with schools. Although community and parent stakeholders did not credit the USI reform with influencing them to become involved in school, they strongly believed that their involvement had a positive impact on student outcomes (see Table 3.1). In fact, 88 percent of the community and parent stakeholders interviewed across the four USI sites indicated that they affected student outcomes in one of three ways: (a) 36 percent stated that they motivate students to finish high school and either begin a career or attend college; (b) 23 percent indicated that they had a positive effect on student achievement scores; and (c) 29 percent responded that they had in general positively influenced student outcomes. These respondents typically indicated that they felt there was room for improvement.

Table 3.1.
Stakeholder Perceptions of the Impact of Involvement on Student Outcomes

Stakeholder Descriptions	Chicago		El Paso		Memphis		Miami-Dade		Total	
	n	%	n	%	n	%	n	%	N	%
Motivate students (complete high school, college, career)	8	15	1	2	4	8	6	11	19	36
Increase student achievement scores/ Improve academic skills	1	2	3	6	2	4	6	11	12	23
Positive impact stated with room for improvement, or no explanation given	2	4	5	10	7	13	1	2	15	29
No affect on student outcomes	0	0	5	10	1	2	0	0	6	12
Total number	11	21	14	28	14	27	13	24	52	100

Motivation

Community and parent stakeholders in our study identified three primary areas in which they assist in motivating students to achieve: (1) completing high school, (2) gaining acceptance to college, and (3) finding a career. Evidence that students who participate in community service and/or learning programs experience gains in social and personal development coincides with the findings of other researchers (Senge, Cambron-McCabe, Lucas, Smith, Dutton, and Kleiner, 2000).

Completing High School. In El Paso and Memphis community and school stakeholders expressed the importance they themselves or their programs place on helping students graduate from high school. One El Paso high school teacher remarked, "The main thing is that we want to help students [by providing support for students who need the extra help] so they do not drop out of school, but finish school" (EP242). According to this teacher, encouraging students to graduate is a school priority. In El Paso the large migrant student population coupled with the high rates of nonnative English speakers creates a challenge for schools to increase graduation rates.

In Memphis a parent described a community program that "encourages students to be in class and be in class on time—knowing that if they're in class they're going to learn." A church leader in Memphis explained why students drop out of school:

> A primary problem is really not aptitude; most of it has to do with motivation. Jesse Jackson said, "you can't beat what you can't see." What that means is that you can never become something that you have never been exposed to. In the community, literally almost half the kids that start off in school drop out for one reason or another, so what we felt was critical was to do something that let the children see the value of education. (ME186)

By involving students in community programs either after school or on weekends, stakeholders believed that they could help motivate students to become actively involved in their education and exposed to experiences that may not be readily available in the students' home community.

Preparing Students for College. In Chicago and Miami-Dade stakeholders from community organizations, universities, and city organizations, as well as teachers, described how they motivate students to apply to college. A community organization in Chicago that provides apprenticeships for high school students interested in construction explained,

If they can get into our adult program which prepares students to get into an apprenticeship or job right away, we consider them actually entering into the program as an outcome. If they get into a state college that offers a manufacturing program or even a construction program, that is a real outcome. (CH176)

This vocational community partner in Chicago encouraged participating students to apply to college vocational programs. This partner indicated that college admission was a superior outcome to obtaining an apprenticeship. A university partner in Chicago responded that the Upward Bound program there has been effective in motivating students to apply to colleges and universities.

In Miami-Dade a school administrator focused on making contact with local colleges and universities. This administrator wanted to provide opportunities for students to see what college was like because the students in the school "at this age and at this point in their lives wanted to see what's next" (MI213). By building relationships with the local colleges and universities, the school administrator was able to provide opportunities for students to visit the colleges. A Miami city organization provided guidance to the students involved in that program on how to "navigate the college application process" (MI214). This Miami-Dade stakeholder recognized the need for additional guidance for students because "there's not that level of help at home and you know a lot of Miami schools are really overcrowded. One guidance counselor will have five hundred students" (MI214). Providing information and assistance to students who may not have been receiving it elsewhere had a positive impact according to this city stakeholder.

Community stakeholders acknowledged the importance of guiding students toward college attendance. Many initiated programs that exposed students to colleges and universities and to familiarize them with the college application process. Without additional support, many stakeholders believed that students would be left behind.

Preparing Students for Careers.   Chicago and Miami-Dade stakeholders were not only vocal about motivating students to apply to college; they were also dedicated to providing opportunities through programs that promoted student attainment of successful careers post-graduation. In Chicago a community stakeholder praised his program for motivating students to begin a career after school:

I think what the School to Work program has done very successfully is broadened horizons for some young people. I have seen that result in light bulbs going on and kids finding a career. They are motivated, at least initially, to do whatever it takes to reach that goal. (CH157)

Similarly, in Miami-Dade a district administrator identified one program in particular, 500 Role Models, as having a huge impact on motivating students to seek professional careers. According to the district administrator,

> Before, these kids all wanted to be pro football players, now they want to become a doctor or technician. They visit the crime lab and they want to get involved in DNA. It's just amazing how they have switched from that sports idea to the more academic, scientific, technological attitude. (MI218)

Motivating students to finish school was particularly important in El Paso and Memphis, whereas encouraging college attendance and beginning a professional career was more emphasized by Chicago and Miami-Dade community stakeholders. Together, these stakeholders expressed their belief that the biggest impact on the students could be achieved by providing encouragement and the motivation to accomplish goals that students may have thought were unattainable.

Increasing Skills and Achievement

It is clear that stakeholder involvement positively affects student skills and achievement scores according to business and city stakeholders. In Chicago the Chicago Bulls professional basketball organization created an after-school program targeting middle schools. According to a Chicago stakeholder, "[They have] concentrated on reading, math, and algebra and the middle schools are doing a lot better than the high schools, really" (CH177).

In El Paso a business that partnered with the El Paso Collaborative and UTEP had its engineers and chemists go into classrooms and do hands-on demonstrations for the kids. One result of this collaboration, according to the business partner, was "the tremendous change and improvement in the science fair projects" (EP236b). A manufacturing business stakeholder in El Paso developed a curriculum that was mathematics intensive. According to this business stakeholder, "They're very high on improving math skills because they clearly understand the benefit to them, as an employer, of having students come out of schools with better math skills" (EP249).

Like the Chicago stakeholders, very few Memphis community stakeholders indicated that their efforts directly impacted student achievement scores or students' academic skills. One city stakeholder described an interactive science program that was supported by a grant and the USI. This program focused on training teachers in grades three through five. To determine the effectiveness of the program, the city stakeholder described research that had been conducted on the program and reported, "The children do seem to be better on standardized tests in science if they have worked with this program" (ME189).

Miami-Dade community stakeholders indicated that they influenced student achievement scores and student's academic skills as much as they motivated students to enter college, start a career, or even finish high school. A Miami-Dade stakeholder explained that there have been "positive student achievement results from USI" (MI216). With the report of positive outcomes on student achievement, involvement from the community and universities increased. As this city stakeholder pointed out, "more people are willing to get on the bandwagon and support the new efforts [once they see evidence of success]" (MI216).

The Importance of School, Community, and Parent Involvement

Parents, schools, and community members saw their involvement as critically important in providing opportunities for students' increased mathematics and science achievement. A Miami-Dade high school assistant principal explained the importance of parental involvement for student success in this way: "If [parents] are not involved, chances are [their] child is probably not getting the best education they could possibly get if [parents] were really involved" (MI210). This principal encouraged parental involvement at his school, recognizing the importance of parents and school staff holding the same goals for students' achievement. Other stakeholders pointed out that while parental involvement is important, the support of community agencies and businesses was critical. An employee of the Chicago Chamber of Commerce, who encourages businesses to provide volunteers for the Chamber's Youth Motivational Program, explained: "Everybody needs to get involved with these students. Not just the schools. Not just the parents. We need people in business and industry; people everywhere need to get involved" (CH158). Many parent and community stakeholders, including a high school counselor in Chicago, shared this sentiment:

> I think stakeholder involvement can really motivate students, make them achieve more if they see a role model. [If students] can sense that someone else is interested in them they are much more likely to achieve well in school and plan for the future (CH160).

Adult role models help students see what they can achieve in their future careers and are an important motivating force for students whose communities lack abundant options for after-school activities and for building future careers. Community and school members as well as parents agreed that individual or whole-business involvement provided examples of individuals who had persisted and achieved educational and career goals.

Most community, school, and parent stakeholders also emphasized the importance of providing places for students to go after school, especially if students participated in academic enrichment programs in these locales. For example, in Memphis a ministry with close ties to one of the high schools in our study maintains a workshop where youths acquire skills repairing old and used computers. A Memphis high school facilitator emphasized the importance of this ministry as an after-school activity: "With the ministry, they have individual representatives that work with our students. They have a computer lab set up in their facility and our students go there in the evening" (ME182). In Chicago a high school program, After School Matters, is aimed at increasing student participation in and community sponsorship of after-school activities. According to the program director:

> [It is important] to have this campaign to create programs for students after school, to get them more interested, to eliminate the idle time . . . because a lot of students we know go home where there is no structure there . . . . (CH167)

Community, school, and parent stakeholders all agree that their involvement in student learning is important because such involvement provides students with access to role models, motivates students, and in turn, promotes student achievement.

Changes Stakeholders Have Seen Since USI Reform

Community and school members mentioned two major changes they had observed at the school level since the inception of the USI: (1) changes in instruction, curriculum, and technology, and (2) changes in school programs and policy. These changes supported the same goal: providing opportunities for students to achieve by enhancing their active learning in and out of school.

Instruction, Curriculum, and Technology.   In commenting on instructional changes in the wake of USI reforms, community, school, and parent stakeholders saw hands-on instruction as a momentous change. An alumnus of a Chicago high school in our study explained:

> It seems like teachers are utilizing newer teaching strategies. I see less seatwork. I see more use of visuals and audio-visual equipment. I see more groupings of students, in terms of doing work. I see more innovative approaches to teaching math and using real-life examples, real-life applications. (CH161b)

While stakeholders from the community and schools commented favorably about hands-on approaches to instruction and the use of real-life applications, as this stakeholder did, others noted that the curriculum itself had changed. An El Paso high school chemistry teacher argued that the curriculum had become more challenging:

> I think the focus and the emphasis on math and science is . . . increasing. I think not only as a district, but nationwide we are seeing the need. And I think that [increased] emphasis [on teaching and learning math and science] is making a big change. Most of our classes . . . the curriculum I think is more challenging for the kids. (EP236A)

Along with increased emphasis on instructional practices and curriculum, stakeholders noted that technology was in wider use. A Memphis high school counselor explained, "[The principal] is a lot into technology. He's enhanced our technological base here" (ME191B). A Memphis youth minister, whose church joined with others in the community to assist local area schools, links the use of computers to instructional practices:

> There's such a fascination with computers. I think just about everybody, everybody who I see in the classroom who's doing it, they, they like the idea of computers. It also gives them some time to kind of work independently of the teacher, without having someone just lecture to 'em. (ME200)

In all four sites, interviewees spoke of these and similar transformations at the school level, including problem solving, hands-on approaches to instruction, and technology aligned with other reforms supporting active learning of more challenging content.

School Programs and Policy.   Community and school members as well as parent stakeholders reported that, in addition to changes in instructional strategies and technology, whole school programs and policies had undergone reform. These included the International Baccalaureate program, Scholars Academy, and Magnet programs, all aimed at providing challenging academic content to replace tracking arrangements that had previously relegated many students to a substandard education. One outcome, as a Miami-Dade high school counselor pointed out, was increased enrollment of higher-achieving students in advanced courses: "Well, I know that there seems to be more of a magnet here where we're attracting calibers of kids that more than three, four, five years ago wouldn't necessarily be coming to this school" (MI229).

In addition to the International Baccalaureate program in Chicago high schools, community, parent, and school stakeholders at each of the four sites favorably mentioned two additional programs: a character-building curriculum that requires students to complete community service hours; and student clustering arrangements that give students a chance to choose a career path early on but also to remain enrolled in high-level coursework required by four-year colleges and universities. Similar programs were mentioned in other districts. As an example of the former, Extended Learning Programs in Memphis were organized at convenient times for students after school, on weekends, and during the summer with an emphasis on service learning in the community to develop social responsibility and commitment. As an example of the latter, in El Paso a high school counselor described a teaching magnet school at one of our participating high schools that was created to address the teacher shortage in El Paso, reframe teacher education by creating a pipeline to the University of Texas at El Paso's (UTEP) teacher education program from the school, and encourage greater "university involvement in the schools" (EP234).

Both academically rigorous schoolwide programs and increased graduation requirements in mathematics and science to prepare students for challenging careers in these fields constituted major policy changes. Changes in graduation requirements often meant a shift from courses such as "general mathematics" and "life science" to four years of rigorous science and mathematics courses—for example, requiring Algebra I in eighth grade. Many teachers and counselors saw this as a positive change. This Chicago high school counselor's statement is representative:

> The requirements that we have here . . . [include] twenty-four credits; this is the first senior class that has twenty-four credits. It's much more intense than it ever was before. I think the kids are responding to that and I think that the colleges have higher expectations as a result of that. (CH160)

However, not all stakeholders saw these changes expanding student options. A Chicago high school stakeholder, who coordinated school-to-work programs, believed these changes encroached upon students' choices and reduced the value of vocational or career coursework:

> One of the significant things that's happened since I started is the number of credits [needed] to graduate from high school in Chicago has increased. And my understanding is that students have fewer electives and more . . . core credits that they have to achieve. So vocational programs get lumped into the elective category and that takes students' freedom away. So that change has, I think, influenced students' decisions about entering the [vocational] program or not. And it's impacted the way teachers . . . teach classes. (CH157)

Although the expansion of challenging coursework to include the enrollment of all students eliminates tracking and other inequities, the debate continues about how best to square students' career interests and skills with academic preparation in anticipation of college and university attendance following high school.

Parents', Schools', and Community Members' Hope for the Future

Major goals for the future cited by parent, school, and community stakeholders centered on increasing school partnerships and involvement. In Chicago a high school employment coordinator of the school-to-work program and a high school director of an after-school program placed a different twist on school and community linkages: "We want to keep linking today's business with tomorrow's workforce. That's what our mission statement is and we want to stick with that, and just kind of keep the program going" (CH158); and "We are going to recruit the community agencies, the businesses. We're going to utilize the university. We're going to do all of those things that will assist us in creating a well-thought-out program" (CH167). The goal of creating and continuing partnerships with community stakeholders to enhance opportunities for students was a hope for the future expressed by many parent, school, and community stakeholders. Some stakeholders explained why they hadn't been as successful in generating partnerships as they would have liked. A university stakeholder, serving as director of Upward Bound and Talent Search and involved with a number of high schools in Chicago, explained:

> We have not cultivated those community relationships as much as I think we should have. Why haven't we cultivated those relationships? My

answer would be because there have not been enough hours in the day to actually do it. I would love to branch out and do a lot of things with the community organizations. (CH174)

Creating partnerships with schools is a time-consuming endeavor. When organizations lack adequate personnel and resources, opportunities to form partnerships diminish. One community stakeholder in Memphis who is involved in Lesson Line, a homework and school information hotline, explained that school district bureaucratic arrangements present an additional problem:

> I continue to see all these barriers when there's no need. I have very strong opinions on what needs to happen in our school system to really change it and I don't know if it's going to happen because of politics. (ME190)

A Miami-Dade community stakeholder involved in education, recreation, and medical services for students in the district predicts a dismal future for partnerships without drastic changes to eliminate resistance to change, expressing similar misgivings:

> I think the best thing that we can do is dismantle our school system—the ways it's currently run. I don't think that you can restructure something when you have a mentality that's been ingrained for twenty years a certain way. Unless you get rid of the people who don't understand what the demands are, you're never gonna have true change. (MI224)

Until barriers identified by these stakeholders have been removed, it is difficult to imagine that the future will increase opportunities to involve parents, schools, and community members in the education of students.

CONCLUSION

Building relationships between the district and the school as well as between the school and its surrounding community can be a daunting task. The Chicago USI at the district level focused on building collaborative arrangements between city museums and the Chicago Public Schools as well as with businesses and community organizations. The El Paso USI established partnerships with local businesses and community organizations before applying to NSF for funding. The

Miami-Dade USI focused its efforts on mass communication to businesses and community organizations to increase knowledge of its reform efforts, with the goal of increasing its interest in participating in the effort. The Memphis USI encouraged existing and new stakeholders to become involved at the level of the school by encouraging school visits and programs. Almost all the district and school administrators interviewed in our study understood the importance of involving the local universities, parents, and community organizations in efforts to enhance student outcomes. However, the majority (70 percent) of the community, university, and parent stakeholders interviewed across the four sites indicated that the USI reform efforts did not impact or had very little impact on their involvement in the schools. While district efforts to involve stakeholders in the USI reform did not appear to have a major effect on stakeholder involvement, both district administrators and stakeholders were aligned in their mission to increase student outcomes in mathematics and science.

4

Professional Development in Systemic Reform

To teach in ways envisioned by the national standards, educators must not only be well grounded in content and pedagogical knowledge; they should also understand how students' cognitions guide their learning of mathematics and science. Teachers must then provide instruction that allows students to be engaged fully and actively with the subject matter at hand. To engage in the best instructional practices:

> Teachers should listen carefully to students' ideas; recognize and respond to student diversity; facilitate and encourage student discussions; model the skills and strategies of scientific inquiry, mathematical problem solving, and technological innovation and ingenuity; and help students cultivate those skills and behaviors. In doing so, teachers should establish a classroom climate that supports learning; encourages respect for the ideas of others; and values curiosity, skepticism, and diverse viewpoints. In addition, teachers should participate in ongoing planning and development of mathematics, science, and technology programs in their schools and seek and promote professional-growth opportunities for themselves and their colleagues. (National Research Council, 2001, pp. 26–27)

In implementing standards-based teaching practices, previous research shows that teachers must have a cognitive handle on both the spirit and the intent of reform rather than simply importing particular activities or manipulatives into the classroom (Briars, 1999). Further, an in-depth understanding of standards-based instructional approaches including curriculum, pedagogical strategies, and assessments is encouraged through extensive, ongoing, site-based professional development. Working with teachers on a long-term basis, often with groups of teachers from the same grade level or department in the same school, is likely to assure that communities of teachers are both formed and equipped to carry out standards-based reforms in their schools.

A major problem in building and sustaining reform at the classroom level, however, is the capacity of the school and district to provide resources to sustain both ongoing professional development and standards-based practices in the classroom. The policy questions addressed in this chapter include: What is the impact on local mathematics and science reform of national, state, and local policies related to systemic changes, including assessments, standards, and professional development? How are resources mobilized among various constituencies such as schools, universities, business and industry, and political agencies? Finally, under what conditions does professional development have the greatest impact?

The NSF's six-driver model described in chapter 1 posits a set of four process and two outcome variables necessary for the successful implementation of systemwide reform of mathematics and science instruction. The first driver calls for "standards-based curriculum and/or instructional materials that are aligned with instruction and assessment." Professional development is the primary vehicle used by school districts for accomplishing this objective. Indeed, NSF has placed considerable emphasis on this aspect of the reform process in both the Systemics and in the more recent and still current Mathematics and Science Partnerships. The Partnership program, like its predecessor the USI, relies on professional development that is planned and carried out by university and school district partnerships. The mission statement of the NSF's Education and Human Resources Directorate, which houses all reform initiatives including the USI and Partnerships, states as one of its goals: "To encourage the development of a cadre of professionally educated and trained teachers to ensure excellence in school education for every student and learner." There is no question that NSF as well as other whole-school reforms view the continuing professional development of teachers, particularly teachers in the middle and high school grades, as the cornerstone of reform initiatives.

In this chapter our focus is on professional development and its affect on classroom practices. Our subsequent analyses presented later in chapter 7 show that professional development that is focused on mathematics and science subject matter, content, and teaching strategies enhances student achievement. In addition, we learned that school culture magnifies the effect of teachers' professional development. What this tells us is that the culture of the school must be developed and nurtured by the principal and teaching staff in order that students are successful and gain maximum benefit from teachers' professional development experiences. Chapter 8 includes analyses of the impact of school culture on a range of teacher and student outcomes.

PROFESSIONAL DEVELOPMENT AND STANDARDS-BASED PRACTICES

Systemic reform as embodied in the NSF's efforts to enhance student outcomes in mathematics and science rests upon intensive and sustained professional development that, ideally, provides teachers with direct, hands-on, problem solving, pedagogical strategies, and content knowledge. The approach taken to professional development in Chicago and El Paso was particularly notable because it emphasized the importance of sustained, long-term teacher support. Researchers and policy analysts recommend a number of strategies for structuring and carrying out effective professional development (Corcoran and Goertz, 1995; Little, 1993; Loucks-Horsley, Hewson, Love, and Stiles, 1998). Most analysts agree that high-quality professional development is characterized by a number of attributes. High-quality professional development:

- Improves student learning.
- Helps educators meet the needs of students who learn in different ways and come from diverse backgrounds.
- Allows enough time for inquiry, reflection, and mentoring and is part of the normal work day.
- Is sustained, rigorous, and adequate to the long-term change of practice.
- Is directed toward teachers' intellectual development and leadership.
- Fosters better subject matter knowledge, greater understanding of learning, and a full appreciation of students' needs.
- Is designed and directed by teachers and includes the best principles of adult learning.
- Balances individual priorities with school and district needs, and advances the profession as a whole.

- Makes the best use of new technologies.
- Is site-based and supportive of a clear vision for student achievement (NEA Foundation for the Improvement of Education, 2003).

In addition to addressing subject matter and pedagogical knowledge, professional development programs should take into account
teachers' prior understanding, provide opportunities to examine instructional practices, and grapple with new strategies and approaches.
Rather than developing these skills in isolation, teachers are most likely
to adopt new approaches to instruction learned during professional
development when professional development activities foster the growth
of school-based communities of learning and practice. Such communities incorporate teams of teachers and are critical both to developing
classroom and school-specific standards-based instructional approaches
in mathematics and science and to increasing student achievement.

Most USI projects employed programs of intensive professional
development focused on modeling constructivist approaches to teaching mathematics and science across grades K–12. However, in the
analyses we conducted of our classroom observations and other data,
the most successful results were apparent for Chicago and El Paso's
programs that (1) assumed a long-term commitment to professional
development on the part of participating teachers; (2) offered both
site-based and off-site activities involving participating museums and
universities; and (3) encouraged the involvement of all science and mathematics teachers in a building as well as principals and other building
staff.

Although professional development programs provided by each
site varied, teachers in all sites had opportunities to attend professional
development programs offered during the school day, after school, or
during the summer. Several sources were used to support professional
development, including the USI; federal, national, and local grants; and
district funds, underscoring the importance of a wide range of resources
to buttress funds provided by a given entity. Sessions at each of the sites
focused on developing content, as well as pedagogical and curricular
knowledge. What varied, as we have just suggested, was the range of
stakeholders involved in both supporting and attending professional
development activities and the duration of these activities. Only two
sites (Chicago and El Paso), one very large and the other mid-sized,
were successful in this regard.

We next turn to an analysis of teacher reports on ways professional
development influenced their classroom practices. Despite some encouraging findings, we found only tenuous links between professional

development and classroom instruction for many teachers. Most teachers seemed to experience a disconnect between their professional development experiences and their day-to-day classroom practices. This is the primary reason we strongly support professional development that is both sustained over time and held frequently in the classroom, where teachers can make a direct link between professional development and their work with students. We begin by discussing teachers' perceptions of reform and their perceptions of how they changed their classroom practices as a result of their involvement in professional development.

Teachers' Experiences of Professional Development

We used a multimethod strategy for examining teachers' perceptions and attitudes about their professional development experiences. In particular, we examined teachers' perceptions of professional development by asking what components benefited them most. We also asked about challenges they experienced as a result of their professional development. In addition, we inquired about whether professional development experiences altered their beliefs about teaching and learning. Finally, we asked teachers about changes in their classroom practices that were attributable to such programs.

Our analyses here are based on two sources: the Survey of Classroom Practices and focus group meetings with teachers conducted at their schools. All teachers completed a Survey of Classroom Practices form that asked them to respond to questions about how they organized and carried out instruction, the time they spent in professional development, the perceived impact of professional development activities on their classroom instruction, and their opinions about major reform tenets. The survey also included sections on assessment practices, influences on their instruction of these practices, and how they prepared for classroom instruction. We examine these and other findings in more detail in chapter 5.

In the first year of our three-year project we conducted focus group meetings at each of the forty-seven schools across the four USI sites participating in this study. Focus groups generally consisted of the five teachers participating in the research over the three-year period, but occasionally also included administrators and other teachers. Participants were asked open-ended questions about their professional development experiences with emphasis on what they liked, disliked, or wanted to see changed. Each focus group meeting was audiotaped and transcribed.

Table 4.1.
Mathematics and Science Teacher Opinions about Student Learning and School Support

	Chicago					El Paso					Memphis					Miami-Dade				
Mathematics	n = 19					n = 22					n = 22					n = 26				
Science	n = 16					n = 12					n = 23					n = 22				
*Topic	0	1	2	3	4	0	1	2	3	4	0	1	2	3	4	0	1	2	3	4
All students can learn challenging content																				
Mathematics	0	11	53	26	26	0	18	14	41	27	0	14	10	62	14	0	27	12	31	31
Science	6	6	19	31	38	0	0	17	42	42	9	13	4	57	17	0	5	14	55	27
Learn basic skills before solving problems or learn basic terms and formulas before concepts																				
Mathematics	5	16	5	42	32	0	14	5	41	41	5	23	0	55	18	4	35	4	23	35
Science	13	38	25	13	13	8	0	25	50	17	0	13	17	52	17	0	23	14	50	14
Learning is better in similar-ability classes																				
Mathematics	6	33	44	17	0	5	23	23	32	18	0	41	9	32	18	4	19	31	23	23
Science	0	40	20	33	7	8	33	33	25	0	0	17	13	48	22	9	18	9	45	18

(continued)

Table 4.1.
(continued)

	Chicago					El Paso					Memphis					Miami-Dade				
Mathematics	n = 19					n = 22					n = 22					n = 26				
Science	n = 16					n = 12					n = 23					n = 22				
*Topic	0	1	2	3	4	0	1	2	3	4	0	1	2	3	4	0	1	2	3	4
Feel supported to try new ideas																				
Mathematics	0	5	21	58	16	0	9	14	41	36	0	5	14	27	55	0	4	23	27	46
Science	0	13	13	31	44	8	0	17	25	50	0	13	9	35	43	0	23	14	36	27
Teachers regularly share ideas and materials																				
Mathematics	0	16	26	53	5	5	18	14	50	14	5	9	9	23	55	4	15	12	31	38
Science	7	27	13	40	13	17	0	42	17	25	4	9	13	35	39	0	27	23	23	27
Teachers actively contribute to curricular decisions																				
Mathematics	11	17	17	50	6	14	23	18	36	9	18	14	23	32	14	12	23	31	27	8
Science	13	33	33	13	4	0	33	42	17	8	9	22	22	43	4	0	36	41	5	18
Have adequate time to work with peers																				
Mathematics	21	37	16	21	5	10	48	5	29	10	32	36	9	18	5	23	31	8	31	8
Science	31	50	13	6	0	17	67	8	0	8	17	48	17	17	0	9	59	18	14	0
Teachers regularly observe each other teaching																				
Mathematics	42	32	16	5	5	23	36	18	18	5	14	27	23	36	0	35	50	15	0	0
Science	20	53	13	7	7	25	67	8	0	0	30	22	4	43	0	23	50	23	5	0

*"Topic" refers to dimensions of student learning & student support. The numbers refer to percentages of teacher opinions.
Percents 0 = % Strongly Disagree; 1 = % Disagree; 2 = % Neutral; 3 = % Agree; 4 = % Strongly Agree.

How Teachers Think about Reform

A belief inherent in standards-based reform is that all students can learn challenging academic mathematics and science course content. There are indications that teachers' belief systems directly affect their practices in the classroom (Fullan, 2001; Stigler and Hiebert, 2000; Peterson, McCarthey, and Elmore, 1996). If teachers have doubts about students' capacity to learn, they are less likely to challenge them with problem-solving activities, laboratory experiments, and other stimulating work.

Our survey results provide some insight into participating teachers' beliefs about student learning (see Table 4.1). Almost 70 percent of mathematics teachers and 77 percent of science teachers agreed or strongly agreed that all students can learn challenging content. However, 70 percent of mathematics teachers and 57 percent of science teachers also *agreed* or *strongly agreed* that students should learn basic skills or terms and formulas before solving problems or learning underlying concepts, and nearly half of the teachers agreed that learning was better in similar-ability classes. Teachers in El Paso and Memphis were more likely to agree that learning basic skills was necessary before solving problems, and teachers in Memphis were also more likely to agree with grouping students by ability than teachers in Miami or Chicago. These findings suggest that most teachers are weighed down with traditional notions of ability grouping and an emphasis on teaching basic skills and facts—all the more reason that teachers need knowledgeable and supportive colleagues to encourage them.

The sine qua non of standards-based school reform is the development of a collegial learning environment at the school level involving teachers engaged in undertaking fundamental change in their classroom practice. More than 70 percent of mathematics and science teachers agreed or strongly agreed that they were encouraged to try out new ideas. Approximately 60 percent thought teachers in their school regularly shared ideas and materials. However, differences in access to power and control over classroom instruction in their schools remain an issue for most of these teachers. Only 31 percent of science teachers and 44 percent of mathematics teachers agreed or strongly agreed that they had opportunities to contribute actively to making curricular decisions in their schools. Teachers in Memphis were most likely to report feeling supported in trying new ideas, sharing materials, and observing other teachers. Teachers in Miami were less likely to feel supported to try new ideas, contribute to curricular decisions, or observe other teachers.

In contrast, more than 60 percent of both mathematics and science teachers reported they did not regularly observe each other and lacked adequate time during the regular school week to work with peers to plan curriculum and instruction. In sum, teachers' responses show that while they felt supported in trying out new ideas associated with reform, they lacked the opportunity to spend time observing other teachers and planning instruction with colleagues. These are important and costly aspects of implementing and sustaining systemic reform at the school and classroom levels. Systemic reform requires an investment of resources to create the necessary conditions at the school level to achieve its goals.

Teachers Talk about Professional Development

During focus group meetings at each of our participating schools, teachers engaged in wide-ranging conversations about topics related to their professional development experiences in mathematics and science. While interacting with their colleagues during our focus group conversations, teachers often spontaneously offered views on a large number of issues. We also asked teachers completing surveys to provide similar information. Although we expected variations among the research sites because of their different sizes and demographic characteristics, most teachers at each location addressed similar topics, touching five broad categories: focus on instructional methods; relevance of professional development to classroom instruction; structure of professional development programs; impact of school-based professional development programs; and improvement of content knowledge.

Focus on Instructional Methods.   Teachers spoke most frequently about pedagogy, the methods of instruction they had learned during professional development focused on mathematics or science. They addressed a wide range of topics, including hands-on activities, curriculum integration, technology in instruction, and group work. For the most part, teachers were positive about their professional development experiences, but also said they desired additional training in specific areas. For example, an elementary teacher in El Paso asserted that her use of hands-on approaches and materials (acquired from professional development sessions) had resulted in students having "become more responsible for [their] own education" (EP2100). Indeed, many teachers clamored for more professional development activities that provided information on the use of hands-on materials for students to use in problem-solving activities. Although teachers acknowledged the benefits of learning instructional practices during the course of professional development, they needed more time to adopt particular strategies in their instruction. One teacher explained, "You just don't internalize enough that you feel comfortable doing the things in the classroom" (CH3200).

Table 4.2.
Time Spent in Professional Development and Impact on Teaching Practice

Activities	Chicago					El Paso					Memphis					Miami-Dade				
Mathematics	n = 18					n = 22					n = 22					n = 26				
Science	n = 16					n = 12					n = 23					n = 22				
	*0	1	2	3	4	0	1	2	3	4	0	1	2	3	4	0	1	2	3	4
Time Spent in Professional Development Focused on In-depth Study of Content																				
Mathematics	0	28	28	22	22	14	14	41	18	14	5	32	23	27	14	20	24	20	20	16
Science	13	27	33	13	13	17	42	17	17	8	14	27	9	36	14	32	23	32	9	5
Time Spent in Professional Development Focused on Methods of Teaching																				
Mathematics	0	39	6	28	28	0	18	45	23	14	0	23	27	36	14	12	24	28	28	8
Science	20	13	27	20	20	18	45	18	9	9	14	24	24	29	10	32	23	32	14	0
	**0	1	2	3		0	1	2	3		0	1	2	3		0	1	2	3	
Impact of Implementing State or National Content Standards on Teaching Practice																				
Mathematics	12	35	29	24		18	18	36	27		10	15	45	30		8	4	67	21	
Science	33	13	33	20		42	25	33	0		4	22	43	30		33	19	33	14	
Impact of Implementing New Curriculum or Instructional Materials on Teaching Practice																				
Mathematics	0	28	56	17		5	14	50	32		10	10	52	29		13	12	50	25	
Science	20	7	40	33		42	17	25	17		9	17	43	30		33	14	38	14	
Impact of New Methods of Teaching on Teaching Practice																				
Mathematics	0	35	47	18		5	23	45	27		5	14	36	45		17	17	50	17	
Science	20	13	40	27		33	17	50	0		17	0	48	35		33	14	38	14	
Impact of In-depth Study of Content on Teaching Practice																				
Mathematics	24	30	41	6		23	41	23	14		10	25	35	30		25	33	38	4	
Science	20	7	60	13		58	17	25	0		22	17	35	26		38	19	33	10	
Impact of Meeting the Needs of all Students on Teaching Practice																				
Mathematics	12	12	71	6		9	18	55	18		5	9	50	36		33	17	33	17	
Science	15	7	50	29		8	42	42	8		17	9	52	22		29	5	62	5	

(continued)

Table 4.2.
(continued)

	Chicago					El Paso					Memphis					Miami-Dade				
Mathematics	n = 18					n = 22					n = 22					n = 26				
Science	n = 16					n = 12					n = 23					n = 22				
Activities	*0	1	2	3	4	0	1	2	3	4	0	1	2	3	4	0	1	2	3	4
Impact of Multiple Strategies for Student Assessment on Teaching Practice																				
Mathematics	6	18	65	12		14	18	41	27		5	0	45	50		30	4	35	30	
Science	13	7	53	27		58	0	25	17		4	9	52	35		38	19	24	19	
Impact of Using Educational Technology on Teaching Practice																				
Mathematics	0	28	56	17		10	32	41	18		10	10	62	19		25	25	33	17	
Science	7	13	47	33		17	33	25	25		10	24	38	29		33	0	48	19	
Impact of Participating in a Teacher Network or Study Group on Improving Teaching																				
Mathematics	41	12	35	12		50	23	18	9		32	9	36	23		63	8	21	8	
Science	56	13	13	19		75	8	17			52	17	30	0		33	5	52	10	
Impact of Participating in a Formal Portfolio Assessment beyond the Classroom on Teaching Practice																				
Mathematics	59	6	29	6		77	9	14			45	23	27	4		88	4	8	0	
Science	53	20	7	20		92	0	8			45	5	41	9		52	19	19	10	
Impact of Attending an Extended Institute or PD Program for Teachers on Teaching Practice																				
Mathematics	53	6	29	12		59	14	14	14		45	10	20	25		67	8	21	4	
Science	60	0	27	13		91	0	0	9		57	0	22	22		76	10	10	5	
Impact of Observing Other Teachers on Teaching Practice																				
Mathematics	72	6	22	0		77	5	18	0		57	5	24	14		75	17	8	0	
Science	69	13	13	6		75	17	8	0		48	17	22	13		62	33	0	5	
Impact of Reading or Contributing to Professional Journals on Teaching Practice																				
Mathematics	28	28	28	17		64	18	18	0		59	5	32	5		50	21	25	4	
Science	38	6	44	13		58	17	25	0		39	13	26	22		48	10	43	0	

Percents *0 = none; 1 = >6 hours; 2 = >6-15 hours; 3 = >16-35 hours; 4 = >35 hours.
** 0 = Did not participate; 1 = % Little impact; 2 = % Trying to use; 3 = % Caused change in teaching practice.

Comments like this reinforce the point that opportunities to try out new strategies with the support of one's colleagues are crucial to reforming mathematics and science classroom practices.

Elementary and middle-grades teachers, in particular, wanted greater integration of topics covered during professional development. Given the scope of the curriculum and the need to address a curriculum that often encompassed many topics over the course of the year, integration of curricular topics is an important means to include related fields in the curriculum across several subject matter areas. During a focus group discussion, an elementary teacher in Chicago observed,

> I would like to see integration, more integration of reading, science into reading, and the integration of mathematics and science together. I would like to see more . . . so teachers won't . . . see it as a separate lesson, but as one whole lesson. (CH1600)

Overall, discussions about pedagogy showed that although teachers recognized the potential of standards-based practices, they were not able to or prepared to readily implement them.

Appropriate use of calculators, computers, and other technologies is an important dimension of systemic reform and one that inspired considerable debate among the participants in our focus group discussions. Many teachers felt unprepared to integrate technology in instruction and to do it well. In fact, some thought it to be an overwhelming experience. "With the new technology and trying to apply it and new methods it leaves very little time for breathing" (EP2400). Teachers in approximately half the focus groups in Memphis and Miami-Dade were particularly frustrated by a lack of access to technology and software after learning during the course of their professional development activities just how effective these tools could be. Others were concerned about the overall availability of the software or equipment that they were learning about.

Relevance of Professional Development to Classroom Instruction. Because attending professional development sessions does not insure adoption of new approaches to teaching, we also asked teachers who took our surveys to evaluate professional development activities they felt influenced their teaching practices (see Table 4.2). Of the twelve professional development activities included on the survey, six were selected by more than half of both mathematics and science teachers as practices they were trying to use or that had caused them to change their teaching practices. These activities included using multiple assessment strategies, using new methods of teaching, implementing new curriculum materials or content standards, using educational technology, and meeting the needs of all students. For the remaining professional development activities the majority of the teachers indicated that they did not participate in the activity or the activity had little or no impact on their teaching. These activities included reading or contributing to journals, participating in networks or study groups, attending an institute for forty hours or more, participating in portfolio assessment activities, and observing other teachers. These results indicate that professional development programs focused upon content, pedagogy, and curriculum, but may not have emphasized the capacity-building aspects of reform required to develop a learning community. This in fact may be a serious omission since it is critically important for teachers to be engaged with their colleagues in carrying out new strategies and approaches based on a strong and shared knowledge base. To sustain standards-based teaching in the classroom, teachers must be able to turn to their colleagues for support as well as be able to rely on their own individual skills and knowledge.

When teachers see tight connections between professional development and their classroom instruction, especially the textbooks they are using, the grade level they are instructing, and their curriculum in general, they can readily make connections to the professional development activities in which they participate. Teachers in a third of all focus groups complained that many professional development activities were not readily applicable to instruction in their classrooms.

These poorly evaluated professional development experiences included professional development content not included in the curriculum, lack of grade-level specificity, and lack of connections to the curriculum (primarily the textbook) teachers used. In three focus groups in El Paso teachers said they "[would like] more training about what we are actually teaching in the classroom" (EP2100). In four focus groups in Memphis, teachers shared that "[we] would enjoy [training if it went] along with the textbook, the activities that go with the different sections of the book" (ME2400). Miami-Dade teachers were particularly concerned about making connections to the text. One expressed a prevailing sentiment when she remarked that she "would like more workshops designed specifically for my grade level and curriculum" (MI1500). Some teachers view the curriculum as the textbook rather than as sets of concepts tied to pedagogical strategies that constitute a repertoire of standards-based approaches cutting across a particular textbook series. When teachers hold this kind of belief, it is very difficult to expect them to change their classroom practices.

We did not hear similar complaints from teachers in Chicago. This may be attributable to the structure of professional development in Chicago. The professional development program in Chicago permitted teachers to select from a menu of sessions provided by the district. As a result, teachers were more likely to select sessions that were of interest to them or that were relevant to their classroom instruction.

Structure of Professional Development Programs. Standards-based instructional approaches emphasize active learning and involvement on the part of students and their teachers. It is not surprising, therefore, that most teachers preferred professional development activities that were more active and supported or encouraged hands-on involvement in training sessions. "Seeing the experiments first-hand was a lot more useful than just reading about it" (CH3100), one teacher commented. However, a few teachers pointed out inconsistencies in the strategy of teachers' assuming student roles during workshops. When professional development coaches suggested role-play or approaches some considered busy work (for example, cutting and pasting), some teachers were highly critical, arguing that these were not engaging and meaningful activities aligned with standards-based practices. At the other extreme, teachers in three focus groups in Memphis reported that "the math workshop was lecture with no involvement from the participants" (MI1500). Teachers from El Paso, Memphis, and Miami-Dade clearly stated that they preferred professional development activities lead by other teachers, reasoning that teachers were better able to engage each other in demonstrating strategies that addressed difficulties they were having.

In our surveys we also asked teachers about the number of hours they spent in professional development during the previous school year. Forty-four percent of the mathematics teachers in our sample reported participating in sixteen or more hours of professional development focusing on pedagogy during the previous year and 38 percent attended workshops that provided in-depth study of mathematics content (see Table 4.2). Thirty percent of the science teachers reported participating in sixteen or more hours of professional development focusing on in-depth study of science content, and 27 percent attended workshops focusing on pedagogy. Across all sites, more than 20 percent of the teachers reported that they did not participate in science professional development. Teachers in Chicago and Memphis reported more participation in mathematics or science professional development than teachers in Miami and El Paso.

Impact of Site-Based Professional Development. Policies that endorse professional development as a strategy for changing classroom practices have great potential, especially if teachers are convinced that the text-book is not solely the curriculum. Classroom practices including instructional strategies, assessments, and the curriculum, as well as textbooks and other materials, have a direct impact on student learning and achievement. However, professional development is most effective when it is carried out as close to the classroom as possible, ideally at the school and classroom levels. Teachers in all four sites found school-based professional development both desirable and effective. Specifically, teachers asserted that mentor-teachers who provided classroom-level support facilitated their ability to incorporate new practices into their instruction. These individuals provided the technical assistance needed to come to terms with new instructional strategies. In cases where mentor-teachers were not available, they were advocated. Teachers in El Paso, Miami-Dade, and Chicago expressed their positions on this subject very clearly:

> I'd give my right arm for somebody to come in and just show me what the heck that they're talking about. I can listen to it. I can conceptualize it but then I come back and try to put it into practice in the classroom and I feel like sometimes we're [not] all on the same wavelength. (EP3100)

> [G]ive us the information, then observe and mentor us to assure under-standing and usage. (MI2300)

> It would be nice if they could send people out and do classroom lessons so we can observe and actually see it in the classroom rather than attend-ing with two hundred teachers. (CH1100)

Site-base professional development activities not only made it easier for teachers to be participants, but such activities also offered the assistance many teachers acknowledged that they needed.

Many organizations, including the National Educational Association (NEA),—the professional organization for teachers,—and the think tank Council for Basic Education exist to assist practitioners in effectively carrying out reform. These groups have protocols and accessible Internet programs to support school-based professional development. For example,

> The Council for Basic Education's Schools around the World and Academy for Teaching Excellence provide an especially rich approach to using student data to improve instruction with an emphasis on mathematics and science achievement in nine countries.[1] These programs offer

teachers a protocol for examining the student work from their own classrooms and for discussing whether or not the work is good enough to
meet high standards. The process guides teachers on how best to inter-
vene with those students who did not achieve. The student work—the
data—provides the evidence for serious discussions about teaching and
learning. Teachers also conduct much of the analysis of instruction and
student work on the Internet with reference to academic achievement
standards. (NEA Foundation for the Improvement of Education, 2003)

The beauty of the Council for Basic Education's programs and others
like them is that they are both readily accessible to teachers and can be
tailored to a particular learning community's needs. During the course of
our research we observed only one school that seemed to have crafted an
approach similar to those we have just described. In this middle school the
principal had arranged for teachers to have access to a large professional
development workshop room staffed full time by an assessment expert
who provided guidance to teachers in interpreting assessment data and
maintaining data files that monitor student progress. Teachers readily
used this facility and valued it highly as a resource.

Content Knowledge.   Although teachers did not address understanding of subject matter in great depth, some reported that they
improved their content knowledge as a result of attending professional
development offerings. In fact, improved subject-matter knowledge
was mentioned in at least one school in each of the four districts.
Typical responses included mentioning that professional development
"[clarified] concepts to be taught" (EP1100).

Level of Reform Implementation and Professional Development Impact on Practice

It makes sense that teachers who work in schools that are most fully
engaged in implementing changes in their instructional practices aligned
with reform would take on more challenging professional development
activities, and this, in fact, was the case. In carrying out our analysis of
teachers' reported hours of professional development involving in-depth
study of mathematics or science content and methods, we learned that
the number of hours focused on methods was much higher for teachers
who worked in high-reform implementation schools as opposed to those
teaching in low-reform implementation schools. These reform-level designations of high-reform, moderate-reform and low-reform implementation schools were assigned by USI district staff based upon the district
staff's view of each school's participation in the USI reform.[2]

Teachers in high-implementation schools reported taking more advanced mathematics courses and more mathematics education courses than teachers in moderate- and low-reform implementation schools. They also mentioned fewer refresher mathematics courses than teachers in both moderate- and low-reform implementation schools. Teachers in moderate- and low-implementation schools, in contrast, were more likely to see professional development activities overall having little or no impact. These teachers were not as likely to be involved in teacher networks or study groups, formal portfolio assessment activity outside the classroom, an institute or professional development program of forty hours or more, or observation of the instruction of other teachers. Science teachers in high-implementation schools reported fewer science content courses than teachers in moderate- or low-implementation schools, although these teachers completed more science methods courses than their counterparts in low-implementation schools.

Whether or not our teachers worked in a school perceived by USI staff as being at a high or a low level of reform implementation appeared to color the experience of the science teachers in our research. First, the total amount of time in the past twelve months spent on professional development or in-service activities providing in-depth study of science content was similar for teachers working in schools at each of the three reform levels. Teachers spent between six and fifteen clock hours in content area professional development. Second, while teachers in schools implementing high and moderate levels of reform spent a similar amount of time in professional development focused on methods of teaching science, between six and fifteen hours, teachers in low-reform schools spent less than six hours in professional development. Teachers in high-implementation schools reported that a number of professional development activities influenced their attempt to use a number of classroom practices, namely, (1) implementing content standards; (2) implementing new curriculum or instructional materials; (3) undertaking new methods of teaching science; (4) engaging in in-depth study of science content; (5) meeting the needs of all students; (6) carrying out multiple strategies for student assessment; (7) implementing educational technology; and (8) reading professional journals.

These findings reveal that teachers in high-reform implementation schools are much more likely to report that they applied what they learned in professional development activities in their classrooms than teachers in moderate- or low-reform implementation schools. Additionally, teachers in high-reform implementation schools were more likely to use educational technology in the classroom.

IMPLICATIONS FOR PRACTICE AND POLICY

We have learned important lessons about professional development from the teachers who participated in our study. First, in order to sustain systemic reform, professional development should be tied to subject matter content *and* pedagogical strategies, incorporating challenging subject matter and instructional approaches. In addition, professional development should occur at the school site, especially when the implementation of new strategies is in question. This provides teachers an opportunity to use what they have learned and enables them to obtain assistance when necessary. This need for site-based professional development is consistent with earlier studies of professional development (Wilson and Berne, 1999). USI-underwritten professional development, however, did not effectively address two critical issues. These are teachers' beliefs about the nature of teaching and learning and the creation of professional learning communities.

Teachers provided little or no evidence that professional development addressed prior assumptions about how students learn and the beliefs underlying those assumptions. Conversations during the course of our focus group meetings suggested that teachers' values and beliefs interfere with the goals of standards-based reform efforts. In the example that follows an El Paso elementary teacher reasons that if drill and repetition worked to correct "careless errors" for her as a student learning mathematics, it should still work in today's classrooms:

> The problem that students are having is the careless errors. And I think it is because of lack of practice in the drill. And not that I am against higher order thinking, but if they don't have the drill, I had the drill as a child, and when I got the higher order thinking, the process fit in. (EP2300)

Just as is the case for teachers who see the textbook as the curriculum, teachers who are committed to drill and practice are unlikely to be receptive to pedagogical strategies focused on standards-based instructional approaches. Further, without examining how their beliefs might limit the opportunities they provide their students, teachers are unlikely to change their usual practices. Indeed, research suggests that professional development activities that address teachers' content knowledge and beliefs can positively influence their instructional practices (Cooney and Shealy, 1997; Franke, Fennema, and Carpenter, 1997; Thompson, 1992). In addition, most teachers we surveyed did not participate in extended institutes or professional development programs and did not engage in networks and study groups. These activities might have provided opportunities for teachers to examine their beliefs and to work with peers to change their teaching in line with standards-based practices. The outcome might have been to provide teachers with more confidence in trying new strategies and engaging in more cognitively challenging and enriched curriculum contents and less reliance on ability grouping and "drill and kill."

How professional development is presented to teachers is extremely important. Should developers model the strategies they wish teachers to try out? The answer might be surprising. First of all, professional development practices, including modeling strategies with teachers taking on the student role, were ubiquitous across each of our four sites. Modeling appropriate pedagogy could provide information about how to undertake specific practices; however, teachers claimed that modeling by itself was insufficient because modeling did not address the realities they experienced with their students in their classrooms. Guskey (1986) argues that teachers are more likely to examine their beliefs and practices if they see professional development activities used successfully with their own students. Working with teachers in their own classrooms allows opportunities for teachers to undertake practical considerations necessary to make real changes in instructional practices. This strategy also provides teachers the time to reflect on their underlying belief structures.

Focus group meetings revealed differences between teachers' generally positive views about professional development offerings and the implementation of newly learned strategies. Teachers remarked that they enjoyed particular professional development sessions because they were called upon to play an active role. Nonetheless, these sessions did not allow teachers to examine their own teaching or to experiment with making instructional changes. In fact, many teachers who were vocal about how much they enjoyed particular sessions also claimed that they were not prepared to implement the presented strategies with their students once they returned to the classroom.

While three focus group meetings in Chicago spent considerable time discussing the development of learning communities as part of their professional development experiences, teachers in most focus groups seemed unaware that to sustain reform it was important to create a community of supportive colleagues within the school, although many principals with whom we spoke knew that this is the case (Adajian, 1996; Hord and Cowan, 1999; Lieberman, 1995; Little, 1982). Further, our survey data showed very small percentages of individuals (less than 25 percent) who reported having changed their practices as a result of participating in a teacher network or study group. In contrast to these findings, many teachers saw value in speaking with other teachers who worked in the same discipline or subject matter area. Workshops provided a vehicle for the exchange of ideas (successful lesson plans, labs, and activities) among teachers who would not otherwise have had the opportunity to share experiences and materials with one another, as several teachers pointed out: "It gave us an opportunity to exchange lesson plans and ideas, what worked, what didn't" (CH2200). "[It] allows us to make those professional connections" (EP2400). "Teachers across the city, whatever subject you're teaching, meet and share activities and ideas" (ME2400). "[It] gave me an opportunity to find out how other teachers in the districts were implementing their programs" (MI1100).

The sites in our study have been in engaged in professional development for at least three years; however, the results from our analysis of classroom observations demonstrate that didactic mathematics and science instruction is the norm in most of the classrooms we studied. Because professional development is the engine that drives the reform effort, our findings suggest that each of these aspects of professional development are important if they are to make a significant impact on the implementation of standards-based instruction.

CONCLUSION

Building from our findings to policy, we argue that professional development should be tied to both subject matter content and pedagogical strategies, incorporating challenging subject matter and instructional approaches. In addition, during the course of the school year, professional development should occur at the school site level, especially when the implementation of new curricula and instruction is in question. These findings and policy recommendations are consistent with many earlier studies of professional development. It is also encouraging from a policy perspective that both the most beneficial professional development and the best practices in mathematics and science were reported by teachers in schools perceived as high implementers by administrators at the district level in each of our four USI sites.

5

Instructional Practices in Mathematics and Science Classrooms

This chapter presents data related to teaching and learning in the mathematics and science classrooms we visited over the course of our research. A fundamental principle of effective systemic reform is that changes must occur in classroom practices, and such changes must enhance student learning. Therefore, a major portion of this research was devoted to gaining insights into classroom practices, particularly evidence related to teaching and learning. The research question guiding this component of the study was: To what degree are science and mathematics being taught in a manner consistent with national mathematics and science education standards? The data, collected during the fall of 1999 and spring of 2000, came from various sources, including classroom observations, teacher and student surveys, and teacher questionnaires. Traingulating the data from multiple sources provided researchers an opportunity to draw conclusions and to compare findings with national Mathematics and Science Standards. Additionally, we made week-long visits to classrooms of a subset of high school mathematics and science teachers in the following school year to study students' level of engagement in their mathematics and science classrooms. We discuss our results from that effort in chapter 6.

During the course of our three-year study we documented instructional practices in 188 mathematics and science classrooms. The classroom lessons we observed ranged from tightly orchestrated performances by teachers who led students through exercises in preparation for high-stakes exams to lessons largely conducted by students in teams or in other configurations where students clearly led others in problem solving or using hands-on activities. The Survey of Classroom Practices[1] (SCP) was administered in 215 mathematics or science classrooms. More than 6,680 elementary, middle, and high school students participated in these surveys, and 163 teachers completed the teacher SCP in mathematics or science that included fifty-three items in common with the student version. The previous chapter discussed teachers' responses to questions on this survey that asked about their professional development experiences. One hundred and fifty-five teachers also responded to our series of questionnaires that asked teachers to describe their views about critical issues related to the reform agenda.

One way to think about the quality of classroom lessons is to consider a scale of best practices as a measure informed by national standards for mathematics and science instruction. Thus, we present our analyses of instructional practices using the standards-based reform agenda and best practices as a gauge. To that end, we carried out a comparison of teachers' and students' views of instructional practices reported in surveys, an analysis of researcher reports of classroom practices garnered from classroom observations, and an analysis of teacher responses to questionnaire items. We hoped that teachers and their students who were in schools actively participating in the USI reforms would report that standards-based practices were the norm. As a result of the reform initiative, we expected students in these schools to report that their teachers often provided opportunities for them to work with their peers on problem solving or investigation-based activities and that they engaged in project work outside the classroom. We anticipated finding less emphasis on completing worksheets that were dependent on rote memorization of material. Our classroom observations in turn were designed to constitute independent reports of instructional practices that corroborated or contested teacher and student reports of classroom activities.

In our analyses we examined differences between teachers' and students' views and found that students are often less sanguine about their teachers' use of standards-based practices than the teachers themselves. To understand these differences in perceptions we conducted a cluster analyses that takes into account the points of view of teachers, students, and researchers. We discuss the results of this analysis in conjunction with case exemplars.

While we focused on teachers who were working to implement standards-based practices that are at the heart of the USI professional development agenda, our observations led us to believe that changing teaching practices is more difficult than simply learning how to implement new teaching methods or implement new content standards. We agree with Stigler and Hiebert (1999), who describe teaching as a cultural activity. They assert that:

> Teaching, like other cultural activities, is learned through informal participation over long periods of time. It is something one learns to do more by growing up in a culture than by studying it formally. People within a culture share a mental picture of what teaching is like. We call this mental picture a script. . . . Cultural scripts are learned implicitly, through observation and participation, and not by deliberate study. This differentiates cultural activities from other activities. It is more like participating in family dinner than learning to use the computer. (p. 86)

Because changing practice requires complex changes in mental models of instruction, changing teaching practices involves far more than simply substituting one set of activities with another set that is modeled for teachers during professional development. Most professional development activities supporting the USI that we observed appeared to be operating under the assumption that participating teachers were being trained in using new activities that would replace old activities they had used in the classroom. As we have argued elsewhere in this volume, in addition to a shift in teachers' mental models of instruction, reform is also contingent upon a strong supportive school culture that includes opportunities to try out new teaching strategies with the active participation of a school-based learning community.

INSTRUCTIONAL PRACTICES

The National Council of Teachers of Mathematics (NCTM) carefully constructed standards that indicate what students should know and be able to do. In conjunction with these curriculum standards, NCTM outlined a vision for classroom practices to enhance the achievement of all students manifest in the process standards. The NCTM's five process standards include: communication as a way of sharing ideas and clarifying understanding; problem solving that engages students in tasks for which methods to derive solutions are not known in advance; representation of processes and products such as diagrams and symbolic expressions; reasoning that encompasses developing ideas, exploring phenomena, justifying results, and using mathematical conjectures in all content areas; and connections among mathematical topics in contexts that relate those topics to other subject areas (NCTM, 2000).

The National Research Council (1996) in the *National Science Education Standards* (NSES) presents a vision of a scientifically literate populace. Like the NCTM mathematics standards, the NSES outlines what students need to know, understand, and be able to do to be scientifically literate at different grade levels. The NSES standards describe an educational system in which all students demonstrate high levels of performance, teachers are empowered to make the decisions essential for effective learning, interlocking communities of teachers and students are focused on learning science, and supportive educational programs and systems nurture achievement.

Although the use of standards-based practices is encouraged, they are not widespread in urban schools. Haberman (1991), among others, is vocal in his critique of the kind of instruction frequently found in urban school settings. He identifies practices that emphasize active teaching and passive student learning through a pedagogy characterized by the teacher:

- Giving information and directions;
- Asking questions, making and reviewing assignments;
- Monitoring seatwork, reviewing for and giving tests;
- Assigning and reviewing homework;
- Settling disputes;
- Marking papers and giving grades.

While such practices as these can and occasionally do occur in standards-based instructional contexts, a pedagogical approach composed solely of these characteristics is what Haberman terms a "pedagogy of poverty" that constitutes what we describe in this chapter as "traditional" or "teacher-centered" classroom practices.

Using the Secondary Teacher Analysis Matrix (STAM) developed by Gallagher and Parker (1995) and authentic teaching variables developed by Newmann and Wehlage (1995, 1996; D'Agostino, 1996), we classified teaching practices into three general categories that describe typical patterns of teacher actions and beliefs: teacher-centered (didactic), subject-centered (conceptual), and student-centered (Simmons et al., 1999).

Newmann and associates (1996) determined that successful reform efforts are characterized by four factors: a strong focus on student learning, authentic pedagogy, the schools' organizational capacity, and the external support the school receives. In these schools, core activities were oriented toward a vision of the intellectual quality of student learning in the form of "authentic student achievement." Authentic achievement is characterized by three criteria: construction of knowledge where students organize, synthesize, interpret, explain, or evaluate information; the use of disciplined inquiry that builds on an established knowledge base, requires in-depth understanding of problems, and produces elaborations on the findings both orally and in writing; and the development of knowledge that has an aesthetic, utilitarian, or personal value beyond the activities of the school.

Unlike traditional approaches to teaching, authentic instruction involves a focus that is primarily student-centered and emphasizes constructivist approaches to teaching and learning. In a student-centered or authentic instructional scenario students negotiate with their teacher an understanding of the important ideas in the lesson based on student input and discussion to develop a deep, comprehensive understanding of concepts. In such instructional milieus we would expect to see the teacher facilitating student-centered communication, social support, and student engagement with the instructional material at hand. In addition, the lesson itself would evince activities, concepts, and ideas that are significant to the instructional topic and that are covered in a manner that enhances student understanding.

Authentic pedagogy supports the construction of knowledge, the use of disciplined inquiry, and the need to have value beyond school by using authentic assessment tasks calling for written work. Authentic instructional practices promote higher-order thinking and deeper, more thorough, exploration, involve substantive conversations among students and teachers, and help students to make connections between substantive knowledge and public problems or personal experiences. Newmann and Wehlage's (1995) work demonstrates that students whose instruction is grounded in authentic pedagogy learn more and that authentic pedagogy boosts achievement for students regardless of race, gender, or family income.

Teacher-centered classroom practices reflect beliefs that the teacher is the primary pathway students must follow to gain content knowledge. Such teachers feel responsible for organizing and delivering content knowledge and stress facts, terminology, and algorithms that primarily depend on rote learning. These teachers tend to be self-absorbed during instruction and so intent on the lesson that they may from time to time be oblivious to the actions and behaviors of their students.

Subject-centered teachers emphasize the descriptive, analytical, and/or explanatory nature of mathematics and/or science by placing subject-matter knowledge as the focus of their teaching rather than pedagogy or student learning. These teachers place content organized around key concepts of the subject at the center of their instruction and emphasize summative assessments. They tend to believe there is a core body of knowledge (composed of mathematical and scientific concepts, facts, and procedures) that all students must learn, and that the curriculum should be constructed around such concepts as they are typically embedded in textbooks.

Finally, in addition to concern about subject matter, *student-centered teachers* are deeply interested in students' ability to use knowledge and to find new information when necessary. In contrast to subject-centered teachers, student-centered teachers develop and implement classroom instruction based on the students' prior understandings and experiences. Student-centered teachers stress the nature of mathematics and science as negotiated understanding, inquiry, and problem solving. Because they are less concerned that everyone shares an identical common core of facts, their assessments are likely to be nontraditional—for example, focusing on students' strategies for solving problems and on using student portfolios and presentations for assessment purposes. These teachers view their role as facilitating learning, and they focus on formative assessments that encourage students to develop responsibility for their own learning.

While a student-centered approach on its face seems most in line with a reform agenda incorporating constructivist approaches, Simmons et al. (1999) describe the possibility of dissonance between an emphasis on learning as negotiated inquiry (characteristic of student-centered teachers) and an emphasis on "fun activities" that do not sufficiently address content or why we "know what we know." It is important to recognize that thinking about students, teachers, and teaching as we do in our clustering analyses results represents a continuum of practices from more teacher-centered practices, shifting to more subject-centered practices, and then to more student-centered practices rather than a set of discreet, tightly bound groups. The case studies that follow the discussion of our cluster analysis results are exemplars of subject-centered, teacher-centered, and student-centered instructional practices.

UNDERSTANDING INSTRUCTIONAL PRACTICES: OUR MEASUREMENT TOOLS

Classroom Observational Tools

To capture a thorough understanding of classroom practices and to do so during a limited time period (usually a single classroom lesson ranging in duration from 40 to 120 minutes), we developed a set of observational coding instruments. Observers recorded the specifics of instructional strategies and technology used during the course of the lesson, noting classroom activities and dialogue pertinent to teaching and learning. Observers were instructed to make note of how students in each observed classroom were configured—whether in small groups or as the whole class. Observers later used their notes to complete a running record of classroom lessons as well as an observation checklist. Observers recorded information that would later allow them to describe at least the following in detail:

- The nature of the activity/activities;
- The logic the teacher used to set up activities;
- Questions the teacher posed;
- Responses the teacher provided to students' questions.

When possible, observers also kept running records of students' comments. What we were aiming for in our field notes included:

- As complete a description as possible of the activity;
- Teachers' and students' activities during the lesson;
- The materials used, including both instructional materials and technology;
- Student grouping arrangements.

The coding instrument that we employed to analyze our observations was derived from the Secondary Teacher Analysis Matrix (Gallagher and Parker, 1995) and authentic teaching practices (Newmann and Wehlage, 1995) described earlier.

Teacher and Student Survey Tools

To identify effective standards-based instruction, a coherent, well-planned instrument must be used to discriminate among learning environments that are reformed, traditional, or somewhere in between (Calhoun, Bohlin, Bohlin, and Tracz, 1997). In this study we relied upon teacher and student responses on a survey of enacted curriculum developed by Rolf Blank, Andrew Porter, and John Smithson (2001), as well as our observations, to determine the degree to which instruction is standards-based and therefore in line with NSF reforms. We used both students' and teachers' responses to items from the Survey of Classroom Practices (SCP) to create a picture of teachers' day-to-day classroom practices. We also used questionnaires that provided our teachers with opportunities to comment on various aspects of the reform agenda.

Teacher and Student Surveys of Classroom Practices

The Teacher Survey of Classroom Practices in Science and the Teacher Survey of Classroom Practices in Mathematics are each composed of 155 items asking teachers to report school and classroom demographics and describe their teaching practices regarding homework, instructional activities, use of instructional time, problem-solving activities, and assessment practices. The Student Surveys of Classroom Practices for mathematics and science were each composed of fifty-three items in common with their teachers' version of their instructional practices. In addition, the teacher surveys asked them to indicate policies and practices influencing their instruction, how well prepared they were to meet teaching goals, how many professional development courses they have taken, the impact of coursework on classroom practices, and, finally, subject area coursework completed as pre-service and in-service teachers.

Teacher Questionnaire

The initial questionnaire was given to participating teachers in the fall of 1999. This and subsequent questionnaire items asked teachers to write about their experiences with the USI; changes in their instructional practices; types of classroom assessments they regularly used; professional development; the effectiveness of their professional development activities; instructional materials they had received from the USI; input received regarding curriculum and instruction; and use of community resources. In addition, five special topic questionnaires—one each month—were faxed to teachers beginning in January 2000. These were designed to solicit teachers' views on a number of issues: the impact of policies, mandates, and student assessments on classroom practices; the impact of professional development on classroom assessment practices; technology access and use; and curriculum decision-making processes. A total of 155 of the teachers from forty-one schools responded to all or parts of our questionnaires. The open-ended questionnaire responses were transcribed and entered into a database for analysis.

Teachers and Their Students

At each of the forty-six USI school sites the five teachers selected for participation were asked to complete the SCP instrument while their students completed the student version for either mathematics or science. Demographic information about our 230 teachers was presented in chapter 1. Demographic information about our students is presented in Table 5.1, which describes characteristics of the 4,681 students who

Table 5.1.
Description of Student Participants in Four Sites

	El Paso		Chicago		Memphis		Miami-Dade	
	n	%	n	%	n	%	n	%
School Level								
Elementary	359	39.4	229	25.4	403	33.5	767˙	46.1
Middle	397	43.5	530	58.8	493	40.9	458	27.5
High	156	17.1	142	15.8	308	25.6	439	26.4
Total Surveys	912		901		1204		1664	
Subject								
Math	446	48.9	491	54.5	573	47.6	895	53.8
Science	466	51.1	410	45.5	631	52.4	769	46.2
Language								
English	624	69.5	757	85.2	1098	92.4	1154	70.6
Gender								
Female	429	48.0	458	51.2	650	55.0	884	54.0
Male	464	52.0	436	48.8	532	45.0	753	46.0
Missing Data	19		7		22		27	
Race/Ethnicity								
Caucasian	147	16.5	22	2.5	107	9.1	208	13.9
African American	55	6.2	661	74.6	976	83.1	647	43.3
Latino	623	70.1	114	12.9	18	1.5	549	36.7
American Indian	22	2.5	14	1.6	19	1.6	19	1.3
Asian	8	0.9	22	2.5	14	1.2	30	2.0
Pacific Islander	8	0.9	2	0.2	9	0.8	28	1.9
Multiple	26	2.9	51	5.8	32	2.8	14	0.9
Missing Data	23		15		29		169	

completed our surveys in 215 classrooms. Here we see that this sample is fairly evenly arrayed across our four sites, with slightly larger numbers of students participating in Miami-Dade. This is also a highly diverse group of students. Not only are there large numbers of ethnically and racially diverse students within and across our research sites; there are also fairly large percentages of students in El Paso and Miami-Dade for whom English is a second language—almost a third in each case.

LOOKING AT INSTRUCTIONAL PRACTICES

Our observations took place in two hundred classrooms in forty-six elementary, middle, and high schools in our four urban sites.

Classrooms and buildings varied in size, age, and condition. One of our high schools in Chicago, for example, was located in an historic African American neighborhood that changed considerably during the three-year period of our research—gentrification brought an influx of families who purchased large condos costing up to a million dollars in nearby high-rise buildings. These families lived alongside people in shabby brownstones, some of which were cut up into small apartments renting for as little as three or four hundred dollars a month. In El Paso most school buildings (and the housing surrounding them) in two of the participating districts were less than ten years old, carved from the desert by an expanding population that now extended many miles from the center city. In sum, not only were the buildings and classrooms that housed the instructional practices we observed highly varied, but the neighborhoods surrounding them were as well. But all were located in places that were among the most impoverished in the United States. Our intent in analyzing data from our observational field notes and checklists was to determine the extent to which the five process standards (communication, problem solving, representations, reasoning, and connections) were evident in teachers' day-to-day work in mathematics and science.

Checklist of Observed Standards-Based Practices

In Table 5.2 we report the results of our analyses of the Checklist of Observed Standards-Based Practices, revealing the occurrence of standards-based practices during our observations. The checklist was designed to rate the extent to which classroom activities addressed each of the five process standards. The checklist is described in Appendix A. Keep in mind that the lower the rating, the less frequent the occurrence of the instructional strategy in question. Overall, we found that teachers were much more apt to simply "provide opportunities" for students to apply mathematical or scientific ideas than they were to "model" appropriate use of mathematical or scientific concepts by, for example, conducting an experiment or demonstrating a problem-solving strategy. Most often teachers were observed implementing practices that minimally addressed mathematics or science standards through simple acts of communication, representation, or drawing connections.

While teachers often used mathematical or scientific language, fashioning representations to help explain concepts or to make connections with previous lessons, other related topics, or real-world tasks, they less frequently implemented standards-based practices in problem solving and reasoning. They seldom went beyond presenting a specific

Table 5.2.
Checklist of Observed Standards-Based Practices by Site

	Overall (n = 183)		Chicago (n = 42)		El Paso (n = 38)		Memphis (n = 57)		Miami-Dade (n = 46)	
	M	SD	M	SD	M	SD	M	SD	M	SD
Communication										
Modeling	1.28	0.96	1.31	0.90	1.16	0.92	0.94	0.97	1.62	0.91
Providing Opportunities	1.26	1.04	1.31	1.06	1.47	1.02	1.12	1.01	1.17	1.03
Guiding	0.66	0.86	0.59	0.80	0.63	0.75	0.65	0.89	0.69	0.92
Problem Solving										
Modeling	0.45	0.67	0.56	0.80	0.28	0.52	0.45	0.64	0.43	0.67
Providing Opportunities	0.81	1.00	1.09	1.15	0.69	1.06	0.90	0.90	0.57	0.89
Guiding	0.41	0.69	0.50	0.80	0.34	0.75	0.51	0.67	0.29	0.64
Representation										
Modeling	1.27	1.04	1.53	1.08	1.09	1.09	1.06	0.97	1.45	1.04
Providing Opportunities	1.31	1.04	1.41	1.07	1.19	0.97	1.31	1.01	1.21	1.09
Guiding	0.72	0.84	0.84	0.72	0.56	0.84	0.71	0.83	0.67	0.87
Reasoning										
Modeling	0.68	0.75	0.91	0.86	0.63	0.61	0.49	0.61	0.74	0.83
Providing Opportunities	0.87	0.93	0.97	1.23	0.78	0.61	0.86	0.89	0.76	0.88
Guiding	0.53	0.74	0.63	0.79	0.41	0.56	0.53	0.70	0.45	0.77
Connections										
Modeling	0.92	0.99	0.94	1.11	1.22	1.07	0.96	1.04	0.64	0.76
Providing Opportunities	0.73	0.91	0.63	1.01	0.78	0.94	0.90	1.04	0.50	0.55
Guiding	0.44	0.64	0.41	0.71	0.19	0.40	0.51	0.73	0.45	0.55
Classroom Displays	0.88	1.03	0.40	0.87	1.00	1.22	1.13	0.94	0.77	1.01

technique or procedure or providing students with opportunities to work alone or in groups to discover solutions to problems. Teachers were most likely, especially at the high school level, simply to present specific techniques or procedures without employing a variety of strategies that students might attempt in solving a problem. Likewise, most teachers failed to encourage experimentation. It was also commonplace for us to see teachers presenting mathematical or scientific procedures

Table 5.3.
Instructional Practices in Elementary, Middle, and High Schools

	Percentages							
	Overall n = 188	Elem. n = 59	Middle n = 77	High n = 52	El Paso n = 40	Chicago n = 44	Memphis n = 59	Miami–Dade n = 45
Communication								
Teacher-Centered	77.8	81.4	82.4	67.3	83.3	76.9	84.6	67.3
Subject-Centered	16.5	11.9	14.7	24.5	11.1	17.9	11.5	24.5
Student-Centered	5.7	6.8	2.9	8.2	5.6	5.1	3.8	8.2
Problem Solving								
Teacher-Centered	70.6	71.4	77.8	60.0	73.1	89.5	50.0	53.3
Subject-Centered	25.0	23.8	18.5	35.0	19.2	10.5	37.5	46.7
Student-Centered	4.4	4.8	3.7	5.0	7.7	0.0	12.5	0.0
Representations								
Teacher-Centered	7.4	9.1	2.9	14.3	0.0	6.7	9.7	10.0
Subject-Centered	76.5	84.8	67.6	78.6	93.3	66.7	71.0	80.0
Student-Centered	16.0	6.1	29.4	7.1	6.7	26.7	19.4	10.0
Reasoning								
Teacher-Centered	69.6	67.1	69.1	74.4	57.7	70.3	70.1	74.5
Subject-Centered	25.4	27.1	25.0	23.3	26.9	24.3	26.9	23.5
Student-Centered	5.0	5.7	5.9	2.3	15.4	5.4	3.0	2.0
Connections								
Teacher-Centered	5.3	3.8	6.7	5.3	0.0	5.9	8.6	0.0
Subject-Centered	78.7	88.5	70.0	78.9	88.9	94.1	62.9	92.9
Student-Centered	16.0	7.7	23.3	15.8	11.1	0.0	28.6	7.1

(continued)

Table 5.3.
(continued)

| | Percentages | | | | | | | |
	Overall n = 188	Elem. n = 59	Middle n = 77	High n = 52	El Paso n = 40	Chicago n = 44	Memphis n = 59	Miami–Dade n = 45
Social Support								
Poor	10.3	6.1	11.6	14.3	6.7	15.4	11.1	6.9
Mixed	29.9	33.3	25.6	33.3	26.7	15.4	29.6	44.8
Positive	50.5	48.5	53.5	47.6	60.0	61.5	44.4	41.4
Strong	9.3	12.1	9.3	4.8	6.7	7.7	14.8	6.9
Engagement								
Inattentive	13.7	6.0	18.3	14.9	10.3	8.8	13.7	20.5
Lethargic	14.9	6.0	9.9	31.9	15.4	14.7	13.7	15.9
On-Task	66.1	82.0	64.8	51.1	71.8	70.6	64.7	59.1
Deeply Engaged	5.4	6.0	7.0	2.1	2.6	5.9	7.8	4.5
*Lesson Coherence**								
Fragmented	9.9	8.9	12.9	6.8	5.9	14.7	11.5	7.1
Mixed Lesson	43.2	37.5	43.6	50.0	29.4	41.2	50.0	47.7
Coherent	11.7	12.5	12.9	9.1	8.8	11.8	13.5	11.9
Review Activities	24.1	23.2	22.6	27.3	32.4	14.7	19.2	31.0
Class Configuration								
Whole Group	50.0	50.6	45.0	57.1	39.4	50.0	51.1	53.5
Small Group	25.4	29.4	27.0	17.5	42.4	26.0	21.3	22.5
Individual	24.6	20.0	28.0	25.4	18.2	24.0	27.7	23.9
Other Observations								
Test Preparation	13.7	16.1	12.9	8.1	5.9	6.1	20.8	17.1
Class Management	32.3	23.2	41.9	21.0	23.5	33.3	37.7	31.7

* Does not include percentage of unclassified activities.

without explaining the logic behind these procedures or providing students with the opportunity to develop reasoning skills to carry them out successfully.

Although there are variations in our checklist data by site as shown in Table 5.2, overall, teachers in El Paso and Memphis were most likely to use standards-based practices, followed by teachers in Chicago and Miami-Dade. Teachers in El Paso and Memphis also were much more likely to use problem-solving activities in their classrooms and also to guide their students in exploring ways to solve problems, discovering answers to complex questions, and demonstrating their reasoning in so doing. It is not surprising to learn that when we examined these data by grade level across sites, elementary teachers provided more standards-based instruction that incorporated the five standards and also gave their students more assistance with problem solving and representations than teachers in middle and high schools.

Authentic Instructional Practices

The checklist data that we have just discussed provided us with information about how well teachers' instructional activities aligned with standards-based instruction; however, these data do not help us understand classroom practices with any detail or richness. Classroom observational data in the form of field notes are the best source of information in that respect. Two trained researchers coded the transcripts of observers' field notes using the Dimensions of Authentic Instructional Practices Coding Matrix (see Appendix A for a description of the matrix components). Table 5.3 summarizes the results of our analyses along nine major dimensions: communication, problem solving, representation, reasoning, connections, social support, student engagement, lesson coherence, and class organization.

In Table 5.3, which describes the results of our classroom observations, we see that the five descriptors are each further subdivided into additional sets of three categories: teacher-centered activities, subject-centered activities, and student-centered activities. This table also presents our measures of the context of the classroom, including social support, student engagement, and lesson coherence, that were coded holistically for each observation, with specific examples coded throughout the lesson. We used a rating scale to indicate the level of implementation for each dimension (described in Appendix A). We next turn to a discussion of each set of findings.

Communication. Teacher talk dominates most classrooms during instruction. Most teachers devote a considerable portion of class time to lecturing or otherwise presenting material in a didactic fashion. Indeed, as shown in Table 5.3, 78 percent of our classroom observations were predominately teacher-centered (didactic). Teachers asked questions that required students to recall facts; furthermore, the lesson content was highly descriptive and there was little emphasis on explaining material in more than one way. In 17 percent of our observations teachers engaged in subject-centered discourse, explaining content of the lesson by emphasizing procedural knowledge and probing for responses; however, there was usually little conversation among students in such classrooms. In fact, teachers who encouraged student-centered discourse were found in less than 6 percent of our observations. On those relatively rare occasions that students were provided the opportunity to converse among themselves about the lesson, these conversations involved sharing ideas to arrive at a deeper understanding of the problem at hand.

Problem Solving and Reasoning. Our results for problem solving and reasoning parallel those for communication that we have just considered. In both the problem-solving and reasoning categories teachers engaged in didactic instruction in 70 percent of our observations, generally standing at the front of the class or at the blackboard or overhead projector. In these classrooms the teacher typically explained the rationale for steps in undertaking problem solving and expected students to apply the same rationale in solving similar problems. A large percentage (90 percent) of Chicago teachers employed rote problem-solving procedures, providing little explanation or rationale and doing so in the context of whole-group class configurations used during 50 percent of our observations. On these occasions, rather than probing for evidence of understanding, teachers expected students simply to recall factual information.

Teachers in Memphis were more likely to explain the rationale for procedures they used and to expect students to justify or explain their work. In 12 percent of observed Memphis classrooms students made, tested, or justified conjectures about problem solving, although their teachers were inclined to emphasize individual work rather than group work. Although 47 percent of teachers in Miami-Dade usually explained the rationale guiding the procedures they used, they did not consistently expect students to justify or explain their work. In addition, Miami-Dade teachers depended on whole-group instruction or individual student work, not investing classroom time in small-group work. Problem solving in El Paso depended primarily on rote procedures; however, because teachers commonly used unstructured group work, students were observed testing and making conjectures about problems more often than was the case in other sites. Standards-based instruction is enhanced by active student participation in problem-solving activities that involve small-group interaction.

Representations and Connections. Slightly more encouraging findings emerge for activities that involve teachers demonstrating concepts, using representations, and making direct or indirect connections to related topics. Elementary school teachers were most likely to demonstrate ways of representing ideas or events being studied in the classroom lesson. Teachers and students also generated a variety of ways for representing ideas and for students to make connections to related topics or real-world events. A third of middle school teachers used a variety of ways of representing ideas or events and encouraged students to construct connections to related topics or the real world. Our high school teachers depended mainly upon subject-centered representations and connections.

Social Support and Engagement. Providing social support and engaging students are both critical if students are to achieve to high standards. Our analysis of social support found that teachers in 65 percent of our observations provided positive support by vocalizing their high expectations for students. In the majority of the classrooms that we observed there was also evidence of high expectations. Teachers often focused on student success; however, they only rarely encouraged students to support one another. Mixed social support prevailed in 24 percent of the observed classrooms. In these instances the teacher praised students occasionally, while at other times students' efforts went unnoticed. We did not find many classrooms either with poor social support or, at the other end of the spectrum, with extremely high levels of mutual trust. Teachers in El Paso and Chicago organized and managed more supportive classrooms, and, not surprisingly, elementary and middle school classrooms were more supportive than high school classrooms. The most worrisome aspect of these findings may be that, although teachers created a supportive learning community in their classrooms, they might not have been providing students with standards-based strategies for learning and showing what they know—approaches that substantially contribute to student academic achievement.

With respect to student engagement, we found encouraging results in the majority of our classrooms. In 74 percent of our observations students were observed as "on-task" most of the time. However, some students seemed to drift off, showing occasional lapses in concentration in 14 percent of the classrooms we observed. Eight percent of the classrooms were characterized by poor student engagement, with many students in the classroom either inattentive or disruptive. In close to a third of our observations classroom management problems that interrupted instructional activities were noted. El Paso was an exception—only 23 percent of the teachers had classroom management problems. Finally, greater difficulties in managing students occurred in middle school classrooms than classrooms in either elementary or high schools.

Lesson Coherence. When topics and concepts covered during the course of a classroom lesson are connected to an overarching theme, the lesson as a whole is likely to allow students to construct a scaffold that supports their acquisition of mathematics and science knowledge and understanding lessons of this type were classified as coherent. "Mixed activity" lessons focused on significant topics, but related concepts were not covered or activities throughout the lesson were not well connected to overarching concepts. Additionally, "mixed activity" lessons covered some key ideas in-depth; nonetheless, lesson coverage was uneven, with related ideas mentioned but not covered in-depth enough to support concept development. Only 12 percent of the lessons were "coherent," where lesson content was presented as a whole by the teacher (often with the active participation of students) and each topic built on another to foster deeper student understanding. Ten percent of the observations revealed classroom lessons that were "fragmented," with lesson material only superficially addressed and with little connective information linking it all together. The researcher using the rubric could not classify another 11 percent of the lessons. These were situations, for example, where the class may have been playing a game or attending an assembly and not engaged in teaching and learning.

We were disappointed to discover that roughly one-quarter (24 percent) of the teachers we observed were engaged in lessons that involved reviewing material previously presented to the class, such as reviewing for classroom or high-stakes tests. More than 30 percent of the observed lessons in El Paso constituted presentations (with some discussion) of previously covered material, while in Memphis and Miami-Dade preparation for high-stakes tests characterized many classroom lessons.

Survey of Classroom Practices

In our analysis of both teacher and student surveys of classroom practices we expected to see a fairly uniform distribution of standards-based practices resulting from the five-year implementation of the USI reform. This, after all, is a systemwide reform that should be working to transform instructional practices in all classrooms in all district schools. We also expected, however, to see some differences in instructional practices and activities by site as a result of block scheduling (resulting in class periods often ninety to 120 minutes in length), the percent of limited English proficient (LEP) students, and variations in teacher professional development. We wondered if survey results would parallel findings from our observations, and they did.

You will recall that the surveys we used were given to each of our participating teachers and their students. We hoped that by collecting these data along with our observational and questionnaire data—all focused on classroom instruction—we would be able to look at classroom teaching and learning processes using multiple lenses. In presenting findings from the SCP analyses we first describe the overall results of our surveys and elaborate with comparisons by mathematics or science and by site. We then discuss selected areas where there were significant differences among sites (for example, use of standards-based teaching practices). Throughout our discussion, we will tie our findings back to our observational data, reminding the reader of important connections between what teachers and students report about what occurs in their classrooms and what we observed.

Teacher Surveys.   The Survey of Classroom Practices (SCP) was used to gather information about instructional practices in the mathematics and science classrooms. This survey inquired about several aspects: instructional time spent on activities in a recent unit; homework; instructional activities; small-group activities; problem-solving or laboratory activities; hands-on or science information activities; technology and equipment; assessment practices; instructional influences; classroom preparation; teacher opinions; professional development experiences; and formal course preparation. (The means and standard deviations of the items composing each section of the teacher survey as well as section means are presented in Tables B.1 and B.2 in Appendix B.)

To describe the overall results for teachers, we make use of the means by subject, grade level, and site. We also have comparable student results for seven of the categories (homework, instructional, small group activities, problem solving/laboratory, hands-on, technology use, and equipment use), and these also will be discussed.

On the whole, responses for each of the seven aspects of instructional activities from the teacher and student SCP are very similar for the 155 mathematics and/or science teachers who responded with their 4,669 students. We first consider results from our analysis of teachers' SCP responses (reported in Appendix B, Table B.1). Teachers emphasized both homework and problem-solving or laboratory activities as components of their practice. They assigned approximately fifteen minutes of homework once or twice a week and counted homework as part of students' grades. During instruction teachers engaged students in a number of activities, including working in pairs or small groups, using hands-on manipulatives, or watching the teacher solve problems or conduct demonstrations. They involved students in these activities during 20 percent of our observations.

Teachers also pointed out that they routinely have students solve word problems from a textbook or worksheet or analyze data to draw inferences and conclusions. When students are working in pairs or small groups, teachers said students are primarily talking about solving problems. These findings support our observations reported earlier in this chapter.

Unfortunately, teachers infrequently used technology or employed the Internet to collect information, nor did they have students use technology to display and analyze information. The use of technology in these ways supports the NSF reform agenda and bolsters student learning because technology allows students to be engaged actively with mathematics and science materials. Instead, teachers spent most instructional time implementing traditional teaching practices. These included having students watch or listen to the teacher, take notes, work individually, and complete computations from a textbook or worksheet. These findings mirror the results of our classroom observations. During our observations, we seldom saw the use of small student groups organized for problem solving or engaged in working with equipment such as graphing calculators in middle schools or high schools. Standards-based instruction encourages the incorporation of technology use by students into classroom lessons.

Factors that Influence Teacher Practice

Assessment Practices. Teachers also responded to a series of questions about their assessment practices; factors that influenced their instruction; how well prepared they felt to deal with a number of general classroom and school circumstances; their opinions on topics regarding curricula, learning, students, colleagues, and parents; and their professional development experiences and formal course preparation also discussed in chapter 4 and reported in Appendix B, Table B.2. With respect to assessment practices, mathematics and science teachers maintained that district tests greatly influenced their instruction. It is important to understand that testing policies varied across sites; some states, notably Texas and Florida in this case, have elaborate systems for sanctioning low-performing schools, while Illinois and Tennessee do not. Therefore, it is not surprising that many Chicago and Memphis mathematics teachers find that district tests positively affect their instruction. This is not the case in either Miami-Dade or El Paso.

Teachers' assessment practices included rather standard fare such as systematic observation of students, short-answer tests, and extended responses to explain or justify answers, but did not embrace the use of performance tasks, presentations, projects, or portfolios. District and state curriculum frameworks or content standards have a positive influence on what they taught according to teachers' own report. Meeting students' special needs and preparing them for the next grade also shaped teachers' approaches to instruction. Perhaps because teachers valued their autonomy in the classroom, many felt that district tests and parents and/or community had little or a somewhat negative influence on their instruction. Later in this chapter we will consider data from teacher questionnaires that clearly show how district- and state-mandated tests appear to many teachers as a double-edged sword—they can constrict instruction, but such tests also can indicate how well students (and their teachers) are doing in meeting state and district performance standards.

Instructional Preparation. Mathematics teachers were highly varied in their appraisals of their own preparation for instruction. Nonetheless, most felt *well prepared* or at least *somewhat well prepared* to teach students from a variety of cultural backgrounds, teach estimation strategies, select or adapt instructional materials, and integrate content with other subjects. Mathematics teachers in Miami-Dade received support from school administrators, while teachers in other sites were neutral or undecided about administrative support. A similar pattern was found in the case of strategies widely recognized as improving instruction, such as regularly observing other teachers and meeting the needs of all students. Miami-Dade science teachers did not subscribe to the belief that students must learn basic terms and formulas before learning underlying concepts and principles. Teachers in Memphis were likely to agree with this belief.

Memphis and Chicago mathematics teachers thought professional development activities influenced their classroom practices, unlike their counterparts in El Paso and Miami-Dade. El Paso science teachers either did not participate or thought that professional development that included activities covering science content standards, in-depth study of science content, and the use of multiple assessment strategies had little impact on their work, while teachers in Memphis and Chicago tried to use these practices in their classrooms. Portfolio assessment in mathematics had little impact on teaching in Memphis and Chicago, and teachers in El Paso and Miami-Dade did not participate in portfolio activities at all.

On the positive side, most teachers reported being well prepared to teach at their assigned grade level, implement instruction that meets mathematics or science standards, and encourage the participation of minorities and females. They felt only somewhat well prepared to teach physically or learning disabled students or limited English proficient students. Most of our teachers strongly agreed that they enjoyed teaching and felt supported to try new ideas, and they also believed that all students could learn challenging academic course content.

Unfortunately, teachers for the most part reported that they were not able to regularly observe each other's classroom teaching, nor did they have adequate time to work with peers. Most teachers indicated that they participated in six or fewer hours of professional development focused on using multiple strategies for student assessment, implementing new instruction, or engaging in new methods of teaching. Most did not participate in professional development activities that allowed observation of other teachers or formally learning how to undertake portfolio assessment. Low levels of professional development coupled with few opportunities to engage in the most innovative kinds of instructional and assessment activities may help to explain why relatively few teachers undertook standards-based, inventive classroom instructional forms, used technology, or employed alternative assessment practices.

Teacher Reports of Instructional Practices by Subject and School Level. Although teachers' low rates of standards-based instructional activities were similar for both mathematics and science teachers, they did differ somewhat in their approaches to instruction. It is important to examine the data in more detail to determine how mathematics and science teachers varied in their approaches to classroom instruction. We accomplished this by considering how teachers in our four urban locations differed in significant ways from each other in their daily teaching practices based on their survey responses. Of the total 147 items on the teacher surveys, only thirty-one mathematics items and twenty-six science items showed statistically significant differences by site using Tukey's studentized range test $p < .05$ (SAS, 2001). The tables with the specific items where there are significant differences are presented in Appendix B, Table B.1.

Looking More Closely: Differences between Mathematics and Science Teachers. Mathematics teachers assigned more homework and used more traditional practices than science teachers. Although mathematics teachers felt better prepared for delivering instruction than their colleagues teaching science, they were less optimistic in their opinions of students and their school contexts. Mathematics teachers received more professional development than science teachers, focused primarily on substantive content. However, they employed fewer problem-solving activities and hands-on activities in their classroom lessons. Mathematics teachers may have received more professional development than their colleagues because students in Chicago, El Paso, Memphis, and

Miami-Dade faced stiff state-mandated tests in mathematics but not in science at the time of our study.

School-Level Differences. In considering results by school level, although elementary teachers use technology less often than their middle or high school colleagues, they are also inclined to use traditional instructional activities with much less frequency. Middle school teachers were more likely than others to employ problem-solving or laboratory activities, small-group work, hands-on activities, and technology. Middle school teachers also reported the highest use of standards-based assessment practices and reported that they were influenced by district curriculum, standards, or state tests in their work. High school teachers reported the greatest use of equipment and traditional practices, but the lowest means for hands-on activities. Considerable attention has been paid to the middle school in recent years for many reasons, but in large part because many are concerned about how student alienation in the middle grades can lead to dropout during this period (Carnegie Council on Adolescent Development, 1996). This may explain some of our findings favoring middle school practices.

Teacher and Student Survey Differences. As we have been documenting consistently throughout the course of this discussion of our survey and observational study results, the conditions of instruction as well as the strategies used by teachers to carry out instruction varied from one site to another. In this section we present teacher and student responses to the SCP to illustrate the differences in instruction by site detailed in Table 5.4.

The use and importance of homework as well as the nature of homework assigned varied across sites. Mathematics teachers assigned homework at least once or twice a week in El Paso, while teachers in Miami-Dade and Chicago assigned homework at least three or four times per week. Science teachers required less homework less often than their mathematics counterparts. Student responses to the frequency of assigned homework in both their mathematics and science classrooms were similar to those of their teachers. While students reported receiving the least amount of homework in El Paso, they indicated receiving more homework than their teachers reported. The content of mathematics and science homework in Memphis and Chicago was more likely to have a standards-based orientation than homework in El Paso or Miami-Dade—for example, writing reports or revising assignments to improve work.

Using hands-on materials or manipulatives and working with hands-on materials to understand concepts were mentioned infrequently, with the exception of Memphis mathematics teachers. Similarly, teachers

Table 5.4.
How Teachers and Students View Classroom Practices

Activities	Chicago (n = 32)		El Paso (n = 29)		Memphis (n = 44)		Miami-Dade (n = 50)		Overall (N = 155)	
	M	SD	M	SD	M	SD	M	SD	M	SD
Homework										
Teachers	1.78	0.51	1.45	0.49	1.65	0.38	1.62	0.44	1.63	0.46
Students	1.72	0.21	1.59	0.38	1.72	0.21	1.71	0.19	1.69	0.26
Instruction										
Teachers	1.52	0.43	1.43	0.39	1.49	0.22	1.44	0.42	1.47	0.37
Students	1.50	0.16	1.47	0.22	1.52	0.14	1.50	0.19	1.50	0.18
Problem-Solving										
Teachers	1.72	0.55	1.58	0.64	1.72	0.45	1.68	0.47	1.68	0.51
Students	1.66	0.19	1.63	0.25	1.60	0.19	1.64	0.24	1.63	0.22
Small-Group										
Teachers	1.48	0.65	1.36	0.60	1.66	0.41	1.49	0.56	1.51	0.55
Students	1.60	0.28	1.58	0.24	1.69	0.21	1.55	0.28	1.61	0.26
Hands-On										
Teachers	1.65	0.60	1.22	0.62	1.76	0.62	1.39	0.59	1.52	0.63
Students	1.48	0.25	1.35	0.36	1.43	0.27	1.36	0.34	1.40	0.31
Technology Use										
Teachers	1.15	0.59	0.89	0.47	1.16	0.69	1.06	0.59	1.07	0.60
Students	1.31	0.23	1.09	0.37	1.23	0.27	1.15	0.28	1.19	0.30
Equipment Use										
Teachers	1.86	0.73	2.01	1.01	2.49	0.95	1.91	0.91	2.09	0.94
Students	1.66	0.47	1.69	0.62	1.77	0.64	1.51	0.59	1.65	0.59
Traditional										
Teachers	1.66	0.43	1.66	0.54	1.54	0.37	1.70	0.48	1.64	0.46
Students	1.87	0.22	1.78	0.34	1.90	0.17	1.88	0.19	1.86	0.23
Standards-Based										
Teachers	1.52	0.45	1.29	0.51	1.65	0.43	1.39	0.48	1.47	0.48
Students	1.43	0.20	1.35	0.25	1.42	0.18	1.34	0.22	1.38	0.21

(with the exception of those in Memphis) rarely allowed students to work in small groups to improve written work, and also infrequently used calculators. These findings are consistent with the observations we discussed previously in this chapter. Teachers in El Paso and Miami-Dade were more likely to describe their instructional practices as traditional than were teachers in either Memphis or Chicago. Teacher and student responses for Chicago and Memphis were similar. Both teachers and their students reported more frequent use of standards-based practices (for example, doing projects outside the classroom or working in pairs or small groups) than their counterparts in either El Paso or Miami-Dade. Conversely, Miami-Dade and El Paso students and teachers were more likely to report very traditionally oriented practices (for example, watching a teacher demonstration, working individually, or reviewing for a test or quiz) than was the case for teachers and students in Chicago and Memphis. Both El Paso and Miami-Dade are in states that are deeply engaged in statewide high-stakes testing, prompting teachers to engage in drill and practice in preparation for these exams.

Discussion of Observation and Survey Results. In reflecting on the results of both our classroom observations (including the checklist data we have discussed) and results of the student and teacher SCP analyses, we can highlight a number of critically important findings. First, although teachers supported students by praising their hard work and monitoring their progress in achieving critical learning goals, many of the mathematics and science teachers participating in this study did not report frequent use of standards-based practices. Middle school teachers were generally an exception to this finding, along with a few teachers in Memphis and Chicago. Second, teachers reported a heavy reliance on homework, and our observations supported this. Homework was often the first item covered in the classrooms we observed and usually the final activity of the lesson. Reviewing material previously covered in class or previously assigned (as in the case of much homework) dominated practices in many classrooms.

Finally, while many teachers maintained that they used labs and demonstrations to teach science, we did not see much evidence of this in our observations or in students' reports of classroom activities. We next turn to the question of how teachers experienced the current climate of mandated tests and other policies that increasingly are finding their way into the classroom and affecting teachers' work.

How District and School Policies Affect Instruction

Instructional practices can be shaped by a variety of school, district, and state policies. In this section we present teacher's thoughts about how such policies affected their classroom instruction. For this analysis we used data from two sources: the SCP and the set of questionnaires we distributed to teachers during the course of our research. In particular, we look at the impact of high-stakes testing, class size, and block scheduling. Last, we present teachers' views on a range of other policies.

Considerable controversy has attended the widespread use of standardized, high-stakes tests, a practice that is now in place in most of the nation's schools. Many argue that high-stakes testing requirements have led to several negative outcomes: the use of class time for drill and practice in anticipation of the test; the dumbing down and narrowing of curriculum and assessment practices; and a relentless focus on basic skills. Most damning is the allegation that these and other related practices lead to a narrowing of students' knowledge and understanding of mathematics and science (McNeil, 2000; Popham, 2001; American Federation of Teaching, 2001).

One critical dimension on which teaching conditions varied was class size. Miami-Dade teachers reported significantly larger class sizes (approximately thirty students) than did El Paso teachers (approximately twenty students), while Chicago classes were only slightly larger (but less than twenty-five) than those taught by mathematics and science teachers in Memphis (just over twenty-six). Science class sizes were similar to mathematics class sizes in El Paso, but larger in the other three sites (see Appendix B, Table B.2).

We see that the number of hours of instruction and the length of classroom periods also varied by site. District policies vary, and block scheduling, year-round school programs, and other programs contribute additional days to the school calendar. Time in school matters and has a strong effect on student outcomes as research on the so-called summer learning effect shows. Students show gains in mathematics and reading/language arts when they remain in school during the summer months (Cooper, 2003).

Course entrance criteria for mathematics vary among sites, with Miami-Dade teachers reporting the greatest control at the school level over student course choice and Memphis teachers having the least discretion. In Miami-Dade achievement scores and teacher recommendations generally determined student placement in courses, while in Memphis students and parents determined student choices. Next we will look more closely at variations in the way teachers responded to our questionnaires.

Teacher Questionnaires. To determine the extent to which teachers participating in this study believed policies had positive, negative, or, indeed, any impact at all on their classroom practices, we framed multiple questionnaires to solicit their points of view. The open-ended nature of the questionnaire allowed teachers an opportunity to respond to us at length about a number of topics, including their views on decision making at their school and district and assessment practices, among others.

The categories of teacher responses from the open-ended questionnaire topics are included in Table 5.5. Responses are arrayed from the most positive to the most negative reflections that teachers provided on school policies. In other words, those responses from our participating teachers that were more positive or affirming are listed at the top of the table while those that are more neutral and more negative are placed further down. Teachers were most enthusiastic about assessment practices and most negative about input from the district and community on school-level decisions.

Assessment practices that teachers controlled or developed and used in their classrooms were viewed most positively, as contrasted with those imposed by states and districts. For example, a Memphis algebra teacher remarked:

> Over the last five years I have changed somewhat in my grading procedures. Instead of *simply giving answers* all the time students must *explain* certain things they must do to solve problems; what would be the best solutions to solving different type problems; they must *draw conclusions from group work*; and *help me grade other groups by using rubric scoring*. (ME2405) (Emphasis added)

In contrast, the perception of state- or district-mandated assessments was generally quite negative, as in the case of this Memphis teacher discussing the district's use of the Terra Nova, a standardized instrument administered annually:

> Terra Nova: this is a state test selected by the board (of education). This test has caused our students to feel worthless. This is true because they are given a Connected Math book to learn from and given a Terra Nova practice book to practice from; however, nothing has helped them to score high on this test. This makes the teachers and the students feel defeated. (ME2403)

Coinciding with teachers' negative views of district and state assessments were their views of other forms of input from sources at both district and community levels, especially those that influenced decisions at the school level. A teacher in Memphis observed:

> My experience with USI [at the district level] makes me believe that [they are] a very independent group of workers. They make decisions for others without true benefits for the student body they represent. (ME1104)

Table 5.5.
Teachers' Views of Policy Impact on Instruction

Opinion	Policies/Areas	Responses (n)	M*	SD
Most Positive	Classroom assessment strategies	133	0.56	0.56
	Changes in assessment practices	130	0.55	0.65
	Input from teachers	114	0.53	0.78
	Effectiveness/usefulness of PD	126	0.52	0.75
	Changed instruction because of USI	133	0.48	0.77
	Input from the principal	120	0.44	0.74
	Technology change	89	0.40	0.88
	Use community resources	116	0.32	0.85
	USI professional development	131	0.26	0.78
	Experience with USI & USI goals	131	0.25	0.93
Positive	Technology access	91	0.22	0.84
	Address student assessment needs	96	0.17	0.80
	Curriculum decision-making process	192	0.15	0.75
	Grading procedures	346	0.14	0.57
	Instructional materials from USI	121	0.12	0.91
	Technology use	246	0.09	0.41
	Faculty assessment issues	94	0.09	0.81
	Input from university partners	103	0.08	0.89
	Curriculum decision-making impact	274	0.07	0.57
Neutral	Policies or mandates	286	0.02	0.78
	Impact of assessments	316	−0.07	0.64
Negative	Input from parents	117	−0.14	0.85
	Input from local businesses	97	−0.21	0.76
	Input from superintendent	97	−0.24	0.69
Most Negative	Input from school board	100	−0.37	0.73

*Where "Positive" = +1, "Negative" = −1, and "Neutral" responses equal zero.

On the other hand, input from colleagues within the school was viewed positively. In addition, teachers truly valued having a voice in district decision making. A Miami-Dade teacher's response is illustrative:

There are many opportunities for shared decision making in our school. We often create committees to deal with curriculum issues and then share the results with others. As a department chairperson, I have the opportunity for input on curriculum issues in the district at district math department chairperson meetings. Several years ago, I served on the district committee that recommended that Algebra 1 be included as a graduation requirement. (MI3202)

While the use of community stakeholders as resources was viewed positively, most teachers did not favorably regard receiving input from the community. In El Paso a strong and influential group at the University of Texas at El Paso directed the USI Collaborative. A teacher in one of the Collaborative's participating schools commented negatively about the limited level of interaction between the university and the school, stating, "Other than hosting USI meetings, [there is] none." Overall, teachers across the four sites believed that community and school interaction was either nonexistent or unsatisfactory. We suspect that there were few organizational structures to allow communication on a more sustained and regular basis since teachers also voiced a positive response to having the community become involved in the school. A Memphis teacher explained:

I believe strongly in community/school/parent involvement in the school system. At our school, we have many organizations that participate on programs, speak, donate, work on different committees for education. They are resources that enhance education to the level of partnership in education. These organizations are helpful in making career choices and decisions in students' daily life situations. (ME1101)

Overall, the majority of teachers' responses to our questionnaires were positive when teachers described their role in school-level decision making, their participation in professional development activities, their access to computer labs or other technology in their schools, and their classroom grading procedures. Most teachers' responses were neutral when they talked about using technology in their own classrooms, their classroom activities, and state standards. Teacher responses were negative when they concerned state and district testing programs, the type of input received from district and community stakeholders, and in decision making at the district and state levels.

COMBINING PERSPECTIVES ON INSTRUCTIONAL PRACTICES USING CLUSTER ANALYSIS

Because we used multiple data sources—observations, surveys, and our questionnaires—we needed a strategy to understand differing perspectives on instructional practices yielded by our analyses. Cluster analysis was used to make sense of our data for students and their teachers responding to our SCP surveys, combined with results of our independently conducted classroom observations. Cluster analysis is a technique that allows the researcher to group cases that are similar on a given number of variables.

The composite variables used to cluster individual teachers were created from major sections or categories of the SCP that were then standardized into category means. For example, ten items that addressed homework were averaged to create the composite homework variable for teacher self-ratings and the mean rating for an individual teacher's students' rating. Homework items addressed time spent on homework, frequency of homework assignments, and types of homework activities completed. Similar composites were developed from other sections of the SCP instrument, and student mean response was contrasted with a given classroom teacher's mean response (see Table 5.6). Ranked pairs of teacher and student scores on the main sections of classroom practice were analyzed using the SAS Cluster Procedure (2001). Each of these clusters contained a number of subclusters, including an occasional single teacher not included in a subcluster with other teachers, thus arraying our classrooms along a continuum from traditional instructional practices to more standards-based instructional practices.

Upon closer examination we determined that the series of subclusters were characterizing three broader distinctive clusters illustrated in Table 5.6. These clusters coincided with the categories that were used earlier in the analysis of classroom observations. The first cluster consisted of those teachers whose self-reported and student category mean ratings were less than zero. These we characterized as *teacher-centered* (more didactic practice). The second cluster included *subject-centered* teachers (more conceptual practice) exhibiting higher means for instructional practices than those reported by their students. The third cluster was seen as more *student-centered* (more constructivist practice), with student category ratings of classroom practice not only consistently higher than their teachers' own reports, but also with the highest means exhibited for each category with the exception of equipment use and use of technology.

Teacher-centered or traditional teachers in our clusters comprise nearly 25 percent of our sample. Their practice is more didactic in nature, and they are less likely to report using hands-on materials, such as manipulatives or technology. Students strongly agree with their teachers that there are in fact very few instances of standards-based practices or the use of hands-on materials and they also report little use of group work. Approximately 42 percent of these teachers teach at the

Table 5.6.
Mathematics Achievement and Teacher and Student Reports of Instructional
Practices by Cluster

| Topics | Cluster | | | | | |
| | Teacher-Centered (n = 25) | | Subject-Centered (n = 48) | | Student-Centered (n = 31) | |
	M	SD	M	SD	M	SD
Mathematics Achievement	−0.19	0.91	−0.01	0.90	0.39	1.01
Instructional Practices						
Homework Activities						
Teacher Scores	−0.30	0.47	0.30	0.61	0.01	0.56
Student Scores	−0.41	0.60	−0.13	0.52	0.58	0.48
Instructional Activities						
Teacher Scores	−0.30	0.31	0.25	0.46	−0.03	0.32
Student Scores	−0.35	0.31	−0.04	0.26	0.32	0.28
Problem-Solving Activities						
Teacher Scores	−0.34	0.48	0.30	0.50	0.06	0.56
Student Scores	−0.43	0.53	−0.08	0.36	0.48	0.45
Small-Group Activities						
Teacher Scores	−0.38	0.41	0.39	0.52	0.04	0.54
Student Scores	−0.49	0.54	0.06	0.42	0.56	0.45
Hands-On Activities						
Teacher Scores	−0.51	0.58	0.26	0.62	0.28	0.66
Student Scores	−0.59	0.68	−0.15	0.54	0.79	0.51
Technology Use						
Teacher Scores	−0.43	0.49	0.41	0.41	−0.06	0.72
Student Scores	−0.31	0.52	−0.02	0.42	0.48	0.68
Equipment Use						
Teacher Scores	−0.34	0.55	0.39	0.59	−0.01	0.72
Student Scores	−0.45	0.69	0.21	0.69	0.13	0.65
Standards-Based Practices						
Teacher Scores	−0.42	0.25	0.34	0.36	0.06	0.50
Student Scores	−0.47	0.37	−0.05	0.32	0.52	0.28

high school level and more than half (58 percent) hold degrees in a field of mathematics or science.

Subject-centered teachers constitute the largest group of teachers (*n* = 48) in our analysis. The majority of teachers in this cluster teach at the middle (56 percent) or high school (29 percent) level and two-thirds have undergraduate degrees in mathematics or science subject areas (as opposed to degrees in mathematics or science education), while a third also have advanced degrees in these areas. These teachers report using technology, working in groups, and using equipment. Their students also affirm that these teachers employ equipment and use some student group work, but they report values below the mean for the other categories of instructional activities, including hands-on, homework, problem solving, and the composite of more standards-based activities. These teachers also reported more participation in professional development activities.

More than half the teachers in the student-centered cluster (n = 31) teach at the elementary school level and are fairly evenly split between holding degrees in either elementary or middle education or in a field other than mathematics or science or multiple fields that include mathematics or science. Student-centered teachers are more likely to report using hands-on materials and working in small groups. They are less likely to report using technology in their classrooms. Their students affirm their teachers' reports with the highest means for using hands-on materials, working in small groups, and using standards-based practices. This cluster is the only group to report that homework is assigned and completed. Most importantly, this is the sole cluster showing above-average mathematics achievement gains over the course of the reform. Although these three cluster groups vary in many respects, as we have just seen, across the three groups these teachers are quite similar with respect to years of teaching experience. Nearly 75 percent in each group has taught nine or more years, many of them in the same school.

Case Examples of Teacher-, Subject-, and Student-Centered Teachers

In this section we provide profiles of teachers whose classroom practices illustrate approaches taken by each teacher cluster. We have selected these teachers from each of the groupings, taking care to provide illustrative cases at each grade level. Our descriptive accounts of these teachers were derived from a variety of sources that include classroom observational data, student and teacher SCP surveys, questionnaire responses, and material provided by the teachers in focus group interviews.

Teacher-Centered Teacher. Allen Thomas[2] is a white male eighth-grade mathematics and science teacher in a kindergarten through eighth grade elementary school in Chicago. This is an inner-city school designated as a high-reform implementation school by USI district staff. He is well qualified and has been teaching mathematics for twenty-seven years and science for twelve years. Thomas has been teaching twelve years at his current school and holds a master's degree in science and a minor in mathematics with several advanced mathematics courses and mathematics education courses. He is certified at the elementary and middle school levels and also in secondary science and teaches "upper grades" mathematics classes on Saturdays as part of the University of Illinois-Chicago Principals' Scholar Program.

Thomas participated in sixteen to thirty-five hours of professional development that focused on methods of teaching mathematics and six to fifteen hours of in-depth study of mathematics content during the 1998–1999 school year. This level of participation is well above the number of hours taken by many teachers in this study. Most teachers reported fewer than six hours of professional development. Thomas described himself as "trying to use" information from his mathematics professional development activities in his classroom, including the reform agenda concepts of "meeting the needs of all students" and "using multiple assessments." Professional development activities that focused on "implementing state or national content standards," "implementing new curriculum," "new methods of teaching," and "using technology" had little influence on his teaching and he did not participate in other "capacity-building" types of activities such as "study groups to improve teaching," "observing other teachers," or "reading professional journals."

Thomas is ambivalent about the effectiveness of his professional development experiences. In his questionnaire responses he told us that he was most influenced by his university course work and said, "[I was] not affected by professional development in regards to math and science. I was affected by classes related to my minor in math and masters in science." When he described his experience with the USI he said, "The experience has been positive regarding my role as math teacher and science coordinator. I found the early inservice meetings at Region offices to be very helpful." He went on to list six workshops that he participated in that were sponsored by the USI. He described the usefulness or effectiveness of these experiences as, "The majority of [professional development] activities were useful in relationship to skills. They could have had more 'hands-on.'" When asked how he has changed his instruction as a result of his USI experiences he said, "I have expanded my lesson plans to a higher skill level. I have increased student workload in algebra and geometry." His answer, however, does not indicate changes he might have made in his teaching practices.

Thomas's classroom is rather large and filled with decorations and projects created by current and past students. There are two doorways that open into the classroom and on the opposite wall there is a low cupboard beneath windows that reach to the ceiling. Hanging from the ceiling are a variety of inflatable animals and a mural of planets is painted on the ceiling. There are bulletin boards between sets of windows that are filled with pictures of graduating classes and science fair posters with student names and grade levels. There are more than twenty student trophies on shelves on both sides of the classroom. The back of the classroom is filled with cages and aquariums that spread out onto the cupboard near the windows. There are also two computers in the back of the room. The front of the room is covered by a chalkboard along with several bulletin boards. He teaches algebra, geometry, and science. The class we observed lasted one hour and twenty minutes as a result of the school's block scheduling program.

In his classroom desks are arranged in six rows; many are empty and class attendance typically ranges between twenty and twenty-five students in a given class. There were nineteen students in the geometry class that we observed; all were African American, and twelve were female. On his SCP survey responses Thomas described his students as having "mixed achievement levels" and that "student ability is the major criterion for entry into this class." Students had notebooks and no textbooks were in sight during our visit. A few of his students had calculators that they used during the class.

Thomas's lesson, delivered mid-October, reviewed content in preparation for a standardized test to be taken in January. He began the lesson by showing polygons he had drawn on the blackboard. He also had a table on the blackboard with the names, number of sides, and degrees of three- to nine-sided polygons as labels that he completed during the lesson. Students calculated the sum of degrees for each polygon using the formula $(n-2) \times 180$, where n represents the number of sides. Students then called out their answers. Students copied this information in their notebooks as Thomas wrote the information in the table. Occasionally he asked a student to come to the board and place information on the chart. The students appeared to be very comfortable as there was some joking between students and teacher. Thomas used available items to represent polygons and later a cylinder such as a sheet of paper and a can. He continued the lesson by moving from problem to problem. He randomly asked questions that involved a variety of geometric concepts: circles, chords, tangents, congruent triangles, quadrilaterals, trapezoids, perimeter, volume, and surface area. He included occasional test hints such as, "Every arc with a radius over five will be over 100. The answers get big. You'll always have two defined places." He ends the class telling students, "About three days of work done today. . . . I'll give you a break tomorrow."

When we compare teacher and student responses on the SCP we gain insight into differing perspectives on classroom practices. Overall, Thomas's responses indicate little use of standards-based practices; instead, he favors more traditional practices such as having students take notes. Student responses support the emphasis on traditional practices with higher means than their teacher reports. Students feel they do at least an hour of homework composed mainly of arithmetic computation three or four times a week that always counts in their final grades rather than having shorter assignments every day as indicated by their teacher. Students indicate they spend the majority of their instructional time watching their teacher demonstrate how to do a procedure or solve a problem. Students agree with their teacher that they solve problems, take notes, work alone, and take tests or quizzes during class. They also agree that they do not use portfolios or hands-on materials, do projects outside of the classroom, or use technology. Students indicate that their problem-solving activities consist mainly of arithmetic problems from their textbooks or worksheets that do not have many "real-world" applications. They agree with their teacher's view that they solve word problems from their textbooks or worksheets and write explanations about solutions some of the time during class.

When describing their small-group activities, students agree with their teacher that they review assignments, problems, or prepare for tests or quizzes some of the time. In addition, students say that they talk about solving problems and completing written assignments from their textbooks or worksheets while in groups. Teachers and students confirm that they do very few hands-on activities, but students think they build more models or charts and they collect less data by counting, observing, or conducting surveys than their teacher reports. Students think they use measuring tools and calculators fewer times during the year than their teacher reports. Thomas and his students report that they use calculators or other technologies to "learn facts or practice procedures," but students also think they are mainly used on tests. They visit a mathematics computer laboratory once a week.

From his SCP results, Thomas's assessment practices indicate a dependence on traditional practices such as short-answer tests. When deciding what to teach, he feels "strongly influenced" by what is needed to prepare students for the next grade level's coursework. His instructional decisions are based on district curriculum frameworks, state and district tests, textbooks, and students' needs. He reported that national standards, parents, or community concerns have "little or no influence" on his instructional practices. However, he later indicated that he feels "very well prepared" to "implement instruction that meets mathematics standards" and is "well prepared" to "involve parents in the mathematics education of their children." On our questionnaire asking what the faculty has done to address student assessment needs, he asserted:

> We are primarily concerned with how tests #1 [Illinois Board's Assessment test IGAP] and #2 [Iowa test of Basic Skills] affect our school. Test #1 relates to state standards and can affect school rating. Test #2 affects whether third, sixth, and eight graders are promoted according to school board rules. We have worked hard regarding extra instruction time and after-school clubs. . . . [We] prepare students in math, reading, and composition. I would predict 20+ hours. . . . We use various books and consumables prepared specifically for test #1 and #2. In addition, we will use copies of old tests, which cover the required areas of study. . . . [These] tests determine the amount of learning for each skill area.

This response indicates that Thomas' classroom practice is greatly affected by the required state and district testing. He believes that these tests dictate how much exposure to skills in each area he needs to provide for students.

Thomas was keenly aware of the curriculum standards guiding his instruction in science. On his questionnaire, he listed several of the standards and described why he thought they were or were not important. Most of his comments regarding standards he believes to be appropriate revolved around his assumption that students needed to be interested in concepts inherent in them. For example, for the measurement standard, "measure length, area, volume, weight, and time accurately in both customary and metric system CFS #1," he argued that, "students can relate to this policy because they know that concepts will be expanded upon in high school. Students also enjoy metric instruction because it relates to science/science fair."

On the other hand, he asserted that the following state geometry standard was not appropriate for his students: "Verify conjectures, both formally and informally, about characteristics of geometric objects. State goal #9." He said that this standard was "too abstract for eighth-grade students. It can cause confusion which leads to a 'turn-off' regarding geometry." His statements indicate his concern for promoting student interest, but they may also suggest reluctance to challenge students' understanding of important concepts.

We gained insight into Thomas' opinions about the reform agenda by comparing his views on reform with other teachers' more traditional responses. Thomas really enjoys teaching mathematics, and believes that "all students can learn challenging mathematics content," but he also thinks that it is important for students to learn basic skills before solving problems and believes that "calculators should be incorporated only after mastery of basic arithmetic facts." In addition, he thought "students learn best in classes with students of similar abilities." His opinions about collegial learning, believing that most teachers contribute actively to making decisions about curriculum, and being supported by school administration were positive. However, he was ambivalent about fellow teachers regularly sharing ideas and materials. He did not feel that he had opportunities to regularly observe other teachers, although he felt supported by colleagues to try new ideas and learn new things.

Thomas is an experienced teacher who relies on traditional practices. He appears to view teaching and student learning to involve a series of skills that are to be presented, practiced, and tested. He describes this modus operandi in a response about assessment strategies: "Tests, quizzes, homework, and math club. I have found that [these] work to improve skills. I have twenty-plus years of experience and this method has positive outcomes for students." He tends to focus on developing content and skills rather than on developing understanding of the subject matter.

Like others in the teacher-centered cluster, he did not rate himself as a user of standards-based practices (1.00 on a 0 to 3 scale), but he also did not rate himself as a user of traditional practices (1.45) as did others in this cluster. However, his students' means on use of traditional practices were among the highest in our sample (2.17). He has ambivalent opinions about the reform agenda. Although he strongly agreed that "all students can learn challenging content," he still supports tracking students into ability groups, learning basic skills before problem solving, and limiting calculator use until after mastery of facts. Overall, Thomas considers himself very well prepared to teach. Thomas works in a high-reform implementation school that seems intent on improving achievement test scores and his teaching practices appear to reflect this narrower focus.

Subject-Centered Teacher.   Carmen Hernandez is a female Latino fourth-grade teacher in El Paso, Texas. Her school was designated as a high-reform implementation school by USI staff in the El Paso Collaborative. She is a well-qualified teacher with eight years of experience. She has a bachelor's degree in elementary education and a master's degree in administration and is certified in elementary education. She has taken several advanced mathematics courses as well as refresher courses in mathematics and mathematics education. She has participated in between six and fifteen hours of professional development focused on in-depth study of mathematics content and on methods of teaching mathematics. She indicated that her in-depth study of mathematics had little impact on her teaching practice, but professional development activities on implementing content standards had caused her to change her approach to classroom instruction. She also did not participate in any professional development activities involving teacher study groups, observing other teachers, or in activities using mathematics journals. Comments from her questionnaire responses help to explain her feelings that in-depth study of mathematics was too theoretical and not helpful:

> Over the past five years, I have seen the math and science areas come forward from the back burner. This was exciting to see since I have always enjoyed math but could not understand how or what I was taught and the "memorizing" I was taught. I have seen technology, science, and math integration with hands-on activities that not only make sense to me, but also are practical. Theory just doesn't cut it.

Although dismissive of theory as impractical, Hernandez also described how she has changed her teaching practice by linking concrete activities to abstract concepts in a developmentally appropriate manner:

> I believe I have made instructional changes. The main change is to take a child through the stages of developmental process, in other words to take the child from the concrete to the abstract. As well as a child may understand a concept, the idea of using manipulatives, real life hands-on, then to picture, then to written explanation, has a profound effect on their internal understanding and connections in real life.

She reported that she is trying to use activities from her professional development experiences on implementing new content standards and curriculum. She was also influenced by activities that emphasized meeting the needs of all students and using multiple strategies for assessing students. Hernandez described her initial professional experiences as very positive, but she did not consider the follow-up sessions worthwhile:

> My experience with the El Paso USI has been through the USI mentors. The mentor is here weekly at grade level meetings. Occasionally, the shared information is helpful. We had initial investigations training

which was very helpful, but the follow-up training was not worthwhile. Some activities have been useful, for example, science fair, hands-on [activities], and Understanding of Math and Science concepts through activities. [There are] quality Web sites to visit in science and math with hands-on activities.

Her school shared a USI mentor with a neighboring school, as other schools we visited in her district in El Paso.

Like the halls in her elementary school, Hernandez's classroom walls are covered with colorful student work and class projects such as pop-up book reviews, large colorful Venn Diagrams, and a Math Investigations board. In her questionnaire responses, Hernandez said, "Students have a project to complete each nine weeks for the theme we are covering." The majority of the twenty students in her class are Latino, except for one Asian boy. She describes the students in her classroom as monolingual—they are exclusively English speakers, although there are several other fourth-grade teachers whose classrooms are bilingual and some are also multi-age, with both third- and fourth-grade students. Like other classrooms that we visited in her school, Hernandez's classroom contains five computers with Internet access. Additionally, every teacher has a laptop provided by the district that he or she can take home. She said, "Students do a lot more research from this mode instead of books. The computer is used to access and reinforce math skills."

The mathematics lesson we observed was the third day of review in preparation for the school's mock TAAS test at the end of the week. Students are working in groups and she asks one student from each group to get the Geoboards (small boards with uniform rows of pegs over which students can place rubber bands to make geometric shapes). Students jump up quickly and she scolds them for acting "uncivilized." She asks the class what they have been studying using the Geoboards and a student answers "fractions." Hernandez told us that, "We no longer use a math textbook and our lessons are focused on manipulative use to gain real understanding of concepts." She also told us that:

> Our district gives a mock TAAS test in the fall. From these scores, students are offered extra classes to help bring up the scores. These classes try to use a wide variety of methods to teach the objectives (manipulatives, games, etc). We prepare daily with practice TAAS questions. Our school offers TAAS Tuesday after school and Math clinics twice a week.

As the lesson continues, she asks students to make "fourths" and reminds them to show their board to their partner, and then they show the class. She uses different student examples in her discussion. The lesson continues and they move on to eighths. They also use their Math Investigations booklets and they draw and color their Geoboard patterns. Some students are having trouble making sure they have equal-sized parts. It appeared to the observer that Hernandez had chosen this activity as an example of what they typically do, but when students got "stuck," she did not seem sure about where to go with the lesson.

Overall, Hernandez is typical of teachers we describe as subject-centered, reporting more use of instructional activities, problem solving, small groups, and use of technology and equipment than their students reported. She reported twice as much use of equipment such as pattern blocks, rulers, and scales than her students reported. Hernandez reports that she assigns fewer homework activities than her students actually mention, a common discrepancy between teachers and their students. Although her students felt they did participate in activities that involved hands-on materials more often than she reported, they reported more use of traditional practices such as taking notes, working individually, and completing written assignments for a textbook or worksheet.

Student-Centered Teacher. Charles Taylor is an experienced African American male science teacher in an inner-city Memphis high school with high levels of student absenteeism but with a principal and several teachers who grew up in the neighborhood who seemed determined to do whatever was necessary to have all students achieve. Memphis USI staff classified the school where he taught as a low-reform implementation school. Taylor was clearly the exception in this case. He has been teaching at his school for the past ten years and has thirty-three years of science teaching experience. Taylor holds both bachelor's and master's degrees in science, with a secondary science education certification. He has taken more than seventeen semester hours in both biological and physical sciences and has six semester hours in both earth-space sciences and science education. Taylor has also participated in more than thirty-five hours of professional development activities focused on in-depth science content and new methods of teaching science. He thought that these activities changed his teaching practices, including prompting him to place more emphasis on meeting the needs of all students, using multiple assessment strategies, and using more technology such as graphing calculators.

Taylor teaches chemistry and physics in a well-equipped laboratory classroom with an adjacent computer room he shares with a neighboring chemistry teacher. The back of his science classroom has eight perimeter laboratory tables and a center table with running water and electrical outlets. Around the room are microscopes and other laboratory equipment. There are posters on the walls highlighting important people in physics and model molecules hanging from the ceiling. There are charts, student projects, and other student work displayed around the room. There is a box full of student-made "canjos" (a banjo made from a can, wood, and string) used to demonstrate the relationship between tension and length in vibrating strings. Students in a shop class and his physics students collaboratively constructed the canjos. On a table next to the door Taylor has placed a stack of physics books, along with dictionaries and other texts that are used as resources rather than having students rely on a single assigned textbook. There are twenty desks in the front of the room, four rows of desks in the center, and six pairs of desks on each side. An overhead projector dominates the center of the room in front of the board.

His classes have between fifteen and twenty students on average. When students enter the class, they place their books on their desks and move to the back of the room to pick up graphing calculators and protractors and begin working on problems at their desks. The class period is ninety minutes long and class meets five days per week. With this schedule, Taylor is able to complete the course in half a thirty-six-week school year. Like the majority of classes at his school, his class is composed mainly of African American students, who are scheduled into his classes based on achievement and/or teacher recommendation. Although Taylor reported that his students are average achievers, he pointed out that ". . . all these kids are pretty quick with physics except for a couple and the students know to help each other." He said that he usually divides the class into two forty-five-minute parts. The first part is for the students to work individually or in informal groups on assigned problems. Taylor said he is "just a facilitator" during the first part of class, walking around the room helping students.

During our observation, he spoke to his students in a soft voice that students then used themselves, creating a very quiet yet busy classroom. Students received a lot of individual assistance from the teacher; he often got down on his knees next to their desks when assisting students so that he could look at them eye to eye rather than towering over them. Although their desks were arranged in rows and the students were given individual assignments, they often conferred with one another about problem solutions, resolving vector problems, for example. Occasionally Taylor addressed the entire class to discuss a critical point he found students struggling with—for example, resolving a thirty-six-mile-per-hour forty-six-degree angle vector into its vertical and horizontal components.

Taylor also engaged in direct instruction, introducing students to new vocabulary—such as the word "equilibrant"—and to additional problems using parallelogram vector solutions. In his discussions he included several descriptions of "real-world" examples of vectors such as a trapeze and guide wires on telephone poles. During the class he made several specific references to types of questions they might expect see on the ACT exam to be taken later that week. He had the students calculate the height of a flagpole using the angle and length of the shadow from the base of the pole because it was one type of question that tended to trip students up on the ACT.

When we compare teacher and student responses on the Survey of Classroom Practices we gain insight into differing perspectives on classroom practices. Taylor reports, for example, that once or twice a week he assigns homework that takes thirty minutes to an hour to complete, and always uses homework in calculating final grades. His students report that they spend fewer minutes doing homework assignments, but work on homework closer to three to four times a week. Taylor reports that his homework assignments require students to use computations to solve problems and for students to improve their own work. His students agree but say they spend more time answering questions from their textbook or worksheets. During classroom instruction, Taylor's students report that they spend much more time listening to teacher explanations than he reports, and they also say they do fewer laboratory investigations. His students do not think that they work on assignments lasting longer than a week or that they organize or display information in tables or graphs as much as Taylor reports. They also do not report using sensors or probes or other equipment as frequently as their teacher says they do. Overall, this class reports less use of standards-based practices and more use of traditional practices than Taylor asserts that he does; nonetheless, the student mean for standards-based practice is one of the highest of all of the classrooms we observed. This suggests that although teachers in this cluster are moving towards standards-based practices, there remains a substantial gap between their instructional practices and the intent of the reform.

Taylor claims he uses more performance or project assessments than traditional assessments and is influenced by curriculum standards and student needs rather than by state or district tests in planning classroom lessons. This may be the result of faculty work using rubrics that he described in a written response concerning faculty decision making on assessments: "[We] worked on Assessment Team Rubrics. . . . Each of us displayed student work with explanations of the assessment. Most of us feel fairly comfortable with the use of rubrics now. We are still feeling uncertain for any and all assessment processes." He also does a daily "think and do microactivity" relating to a TCAP process that lends credence to the science task at hand. Taylor wrote, "I was not so surprised at the implementation of the 'standards.' [They had] a major impact, for it proved my teaching direction was on task. [I was] much relieved by having taken the 'guesswork' out of the teaching in my areas—chemistry and physics."

Taylor enjoys teaching and reports being very well prepared to teach science and well prepared to teach the students in his school. He has strong positive opinions about all students learning challenging curriculum, yet believes that ability grouping is still important. He said, "All teachers have made themselves available to the students at varying times of the school day. Now it is common practice to meet with students before and after school." Taylor also has strong positive opinions about collegial relations in his school and says that he feels supported by colleagues and administrators. We observed a close working relationship with the chemistry teacher with whom he shares the computer lab. He also believes he has adequate curricular resources, but does not think he has sufficient time during the week to work with peers on science curriculum.

Our analysis of the research team's observations of Taylor's classroom yielded the highest ratings of all of the teachers we observed. Analysis of our observation notes revealed that his lesson was coherent, although it was not clear from this observation that key concepts were appropriately covered in depth, or that the content of the example problems was necessarily sequenced to foster deeper understanding since it was a continuation of unobserved lessons. However, he was masterful in modeling appropriate use of scientific concepts, and also provided students with opportunities to practice using those concepts, as well as providing guidance in their use. He gave students opportunities to attempt to solve problems using a variety of strategies and encouraged different approaches. During our observations, students were not actively encouraged to go beyond the problem-solving rationale or to make and test conjectures about solutions on their own. His practice of spending a good deal of his class time helping individual students may have obscured these types of activities during our observation. Students were expected to use representations, such as diagrams, to assist with their problem solving, but identifying possible limits or exceptions was not openly discussed. He made several connections to mathematical topics and used several real-world references, but students exhibited some difficulties making similar connections on their own.

Taylor cared about developing his students' understanding of science, achievement, and future college success, and it appeared that his students understood his views. Although during our observation he did not appear to actively encourage student group work or conversations about the lesson material, his students frequently shared ideas to arrive at a deeper understanding of the lesson content. Perhaps this was a carry-over from their laboratory activities. Students were deeply engaged and appeared to be following well-established class routines that focused their attention on learning and achieving. There was evidence of positive social support along with good student-teacher rapport. Taylor had high expectations for learning and trying hard, and he focused on student successes rather than dwelling on their mistakes.

Taylor is an experienced teacher who cares about his students' current and future successes. These factors no doubt contribute to the high levels of student engagement that we observed. His responses to questions about the impact of his professional development experiences reveal a teacher trying to move from a long-established personal background of subject-centered teaching experiences to a view incorporating the use of rubrics and the need to use a variety of assessments, along with greater use of technology, and providing all students with increased opportunities to learn challenging science content.

Taylor asserts that his participation in the USI professional development program has greatly influenced the way he currently teaches. Students do not rate him as highly in the use of standards-based practices as he rates himself. Students may believe he uses traditional practices because he has not developed more active group structures that foster student-to-student interactions and create a stronger learning community. This suggests that he is still transitioning from a subject-centered focus to a more student-centered focus. This tension is manifest in his expressed agreement with ability grouping practices and his belief that all students can learn challenging content. However, it is Taylor's strong teaching and academic background, his positive views about collegial relationships, the high level of student engagement in his classes, and his desire to find more effective ways to help his students that distinguish him from his more subject-centered colleagues. It appears that he is not just teaching physics or chemistry; he is teaching his students to learn physics or chemistry so they will have the opportunity to succeed in college and life, and this grants him a more student-centered focus.

CONCLUSION

Our analysis of our observations of classroom lessons using our Authentic Instructional Practices rubric leads us to conclude that, taken as a whole, teachers are providing instruction that does little to promote deep student engagement. In each of the five standards (of communication, representation, problem solving, reasoning, and connections), instruction was an assemblage of 70 percent or more of teacher or subject-centered activities and less than 20 percent student-centered activities. The more common activities observed during our classroom visits involved teachers asking questions that required students to recall facts about lesson content that was superficially descriptive with little emphasis on explaining material in more than one way. Teachers typically explained the rationale for steps undertaken in problem solving and expected students to apply that rationale in solving similar problems.

Analysis of our observations using the Checklist of Observed Standards-Based Practice provides us with information about the level of implementation of the standards. Overall, teachers minimally implemented standards-based practices in communication, problem solving, representation, reasoning, and connections. The highest means (1.3) were for the use of communication and representation. These categories of use for each standard are "modeling activities," "providing opportunities for practice," and "providing guidance" to students. Each category was evaluated for using the activities minimally, moderately, or for the majority of the observation. Means for each of these categories of use in problem solving, reasoning, and connections were all less than one. Teachers were most often found "providing opportunities" for students to use the standards with the lowest means for each standard for "modeling" standards and for "guiding" students in the use of the standards. This suggests that teachers are not yet proficient in implementing these critical aspects of standards-based instruction. Our teachers in El Paso were more likely than others to provide their students with moderate use of visual representations such as graphs, tables, or models of mathematical or scientific concepts to help explain these ideas. El Paso teachers also made moderate use of opportunities to develop reasoning skills, such as asking students to explain the procedure they used to solve a problem. Memphis teachers were more likely to provide students with connections between various topics within mathematics or science, or between mathematics or science and other disciplines. Teachers in Miami-Dade and Chicago were more likely to communicate mathematical or scientific ideas using oral, written, or visual forms, such as explaining the process by which they reached a solution to a mathematical or scientific problem. We found that elementary teachers provided more assistance than our teachers in middle and high schools with guiding students in exploring ways to solve problems and guiding students in their use of mathematical or scientific representations.

In our analysis of our observations of classroom lessons using our Authentic Classroom Dynamics Matrix (see Appendix A, page 221), we found that 50 percent of our observed classroom activities, were whole-group activities, with individual work and group work each using a quarter of the remaining time. El Paso was the exception, with a predominance of group activities in 42 percent of our observations. Only 12 percent of lessons we observed were evaluated as coherent. Most of the lessons we observed were mixtures of activities (43 percent) that were missing important ideas or concepts necessary to develop student understanding. More than 10 percent of lessons in Chicago and Memphis were evaluated as fragmented. Almost a quarter of the lessons we observed were review lessons, with 14 percent tied to high-stakes tests. In El Paso and Miami-Dade more than 30 percent of the observed lessons involved review.

We found students generally on-task in 66 percent of our observations, but lethargic or inattentive in 29 percent of our observations, while only 5 percent were seen as deeply engaged. In 47 percent of our high school observations students were not on-task during our observations, while 82 percent of elementary classrooms and 65 percent of middle school classrooms were found to be on-task. The social support we observed paralleled our student engagement results, with 50 percent of our teachers creating classrooms with positive social support. In El Paso and Chicago the social support was positive in more than 60 percent of our observations, paralleling the on-task engagement in more than 70 percent of those classrooms. In Memphis the results were somewhat mixed. We found the highest social support in 15 percent of our Memphis observations, but less than positive social support in another 41 percent of our observations. There was off-task behavior in 27 percent of our Memphis observations; in Miami-Dade more than 50 percent of the classrooms had less than positive social support, and 37 percent had off-task behaviors.

In our analysis of the SCP for teachers we generally found that reported means were higher for classroom activities using equipment such as measuring devices and manipulatives and successively decreased for other activities such as problem solving and homework. During our observations we did not see much use of equipment. On the other hand, from the SCP responses for students we deduced that means were higher for homework and decreased for equipment use and problem solving. In our observations we found that activities related to homework usually played an important role. Both teachers and students rated technology use lowest across all sites, indicating that use of computers, calculators, and other technology is not as strongly integrated into classroom practices as we had anticipated. During our observations it was common to see two or three computers in each classroom that were not turned on, typically surrounded by stacks of class materials, indicating little routine use.

Both teachers and their students in Memphis by and large reported higher SCP means than their counterparts in our other sites for equipment use, hands-on, and small-group activities. The SCP means for Chicago teachers and students were less than those in Memphis, while means for teachers and students in Miami-Dade were less than means in Chicago. The means for El Paso's teachers and their students were the lowest of our sites except for equipment use. When we looked at the differences between means for teachers and their students, teachers reported more use of equipment than their students. When we looked at teacher and student differences on items indicative of traditional practices, our teachers reported lower means for traditional practices than their students reported. Conversely, teachers reported high means for standards-based practices, while their students reported lower means. Teachers in Chicago reported the highest means for standards-based activities and the lowest means for traditional activities, while teachers in Miami-Dade reported the highest means for traditional activities. In Miami-Dade and El Paso we found the smallest differences between teacher and student means. There were fewer variations among student means across our sites than there were among their teachers. Students' reports of the major role of more traditional practices in instruction across all sites were generally more aligned with the results of our classroom observations than teachers' reports of practice.

Combining our measures of instructional practices, standards-based practices, classroom dynamics, and SCP results aids our understanding of classrooms. For example, the SCP scores in El Paso were conservative, the lowest for both teachers and their students of all our sites. Teachers in El Paso were observed to have a more subject-centered approach to their instruction. During our observations in El Paso, we found more teachers using small-group activities, more students engaged (on-task), and fewer problems with classroom management than in other sites. These positive classroom conditions almost certainly contributed to El Paso's higher standards-based checklist scores because students had increased opportunities to learn, and also contributed to their higher student-centered scores even though teachers mainly depended upon subject-centered instruction.

On the other hand, in Memphis SCP scores for both teachers and students were the highest of all our sites and Memphis had the highest percentages of classrooms with student-centered practices. In Memphis, we found the highest percentage of classrooms with strong social support, deeply engaged students, and coherent lessons when compared to the other districts. Interestingly, Memphis also had a large percent of classrooms with management problems, high occurrence of independent seatwork, and poor social support. These wide-ranging classroom conditions almost certainly contributed to the students' high rating for use of traditional practices and to lower means for standards-based checklist scores in Memphis. To make sense of this complex mix of teacher and student SCP results, our classroom observations, and our standards-based checklist results, we employed cluster analysis.

Our cluster analyses reveal that teachers are at various stages regarding their approval and acceptance of standards-based teaching. In our sample of teachers there is an indication that there is a progression of changing instructional practices from traditional to more standards-based practice, but this progression is slow. The largest group of teachers characterized as subject-centered suggests that progress has been made, but much work is still needed in order to accomplish the goals of the USI. These teachers appear to feel that they are more standards-based in their instructional practices than their students believe or than we observed in their classrooms. Our research results emphasize the importance of soliciting multiple views regarding classroom practices. Reliance on one or even two sources may not provide the depth of information needed to effectively understand what is happening in mathematics and science classrooms. We could argue that many of our teachers' self-reports appear skewed toward the goals of the reform, thereby presenting a result that is more in line with mathematics and science standards. We also could argue that the students' survey responses may not be based upon a real understanding of the reform agenda and thereby are a less accurate indicator of reform implementation. Our one-period "snapshot" of classroom practices could be just a "show" by the teacher and not an indicator of typical practices. We could even argue that district administrators had difficulty accurately classifying schools into "high-reform implementation" or "low-reform implementation" schools. However, by combining these differing views of classroom practices we have more insight as to what constitutes classroom practices and where they align with mathematics and science standards.

The goals of USI were to provide professional development opportunities for teachers to learn to teach in new ways and to improve substantially the learning opportunities for all students. To achieve these goals, professional development activities cannot rely on the conventional staff development practices of disseminating activities with the view that these activities can be substituted into the teacher's current practice. These conventional practices view teaching as a collection of activities and not as a culturally coherent set of activities and beliefs that are bound together.

6

Student Engagement in Mathematics and Science

Our evaluation of the impact of systemic reform moved beyond achievement scores on standardized assessments in mathematics and science to include a measure of students' participation in their own learning. Standards-based reform practices in both mathematics and science call for increased activity that engages students and leads to increased achievement (AAAS, 1993; NRC, 1996; NCTM, 2000). If students are not engaged in their mathematics and science lessons, they are not likely to learn. This chapter is about what motivates students to learn in mathematics and science classrooms and what engages and sustains their attention. Student engagement is directly related to teachers' relative success in implementing the reform agenda and improving teaching practices. We are concerned with important features of classroom interactions—those we think have bearing on how much and what students actively learn in their mathematics and science classrooms. We were guided by our hypothesis that students who are engaged in their classroom lessons are more likely to be learning actively than students who are not so engaged.

Because we wished to examine student outcomes using a measure other than student performance in mathematics on high-stakes tests, we undertook a study of high school students' engagement in mathematics and science classroom activities in a variety of courses (for example, biology, chemistry, physics, algebra II, pre-calculus, and geometry—regular and honors). Specifically, we assessed the impact of classroom activities on high school students' engagement in mathematics and science, in addition to students' thoughts and feelings about their classroom learning using the Experience Sampling Method (ESM) discussed in detail later.

In addition to describing students' classroom experiences, we examined the relationship between the context of the mathematics or science lesson and student engagement. We wanted to know which features of teaching practices (such as organizational or instructional arrangements) encouraged students to be more engaged in classroom learning. We hypothesized that standards-based practices, such as group work and hands-on student activities, would engage students more than traditional teaching practices. We also evaluated the impact of teachers' reported professional development activities on their classroom practices and identified whether students experienced higher levels of engagement in classrooms where their teachers reported that their professional development experiences impacted their classroom practices.

Studies based on large national and international data sets, such as the Third International Mathematics and Science Study (TIMSS), report that U.S. classroom lessons were likely to be characterized by review and repetition of less rigorous and challenging mathematics and science curriculum than was the case in other industrialized societies such as Japan and Germany (Stigler and Hiebert, 1999). Using similar data sets, Yair (2000) and Shernoff (2001) reported that U.S. students were often not challenged by classroom instruction. Following their work, we set out to evaluate classroom practices from the students' perspective, including whether students found classroom lessons repetitive, unchallenging, or, in contrast, relevant to their lives outside school. We examined classrooms involved in a standards-based reform initiative, the USI, that sought to enhance teaching practices and challenge students.

We focus on three major themes that are important for policy considerations regarding students' engagement. First, we describe types of classroom organizational patterns, such as whole-group lecturing and small-group work, and their effect on the level of student engagement. Second, we examine students' perceptions of classroom instruction, such as the relevance of what was being taught to their everyday lives, college interests, job expectations, and future tests. The issue of teachers' use of repetitive rounds of review is also explored. Finally, we look at individual teachers as units of analysis and examine data about their experience with USI reforms.

EXPERIENCE SAMPLING METHOD

In order to probe how students perceived classroom events, we employed a data collection method called the Experience Sampling Method (ESM) (Larson and Csikszentmihalyi, 1983). At each participating high school in our four sites, ten students in each participating mathematics and science classroom (16 teachers and 345 students) were asked to complete a short questionnaire when given several signals by vibrating "beepers" during the instructional period. Beeps were signaled at ten-minute intervals and the beeping schedule was set such that each student received a signal every twenty minutes. Because we followed the same students in the same mathematics or science classes for one entire week, we generally have ten sets of observation points or "beeps" from each student participant.

Students were signaled intermittently and were asked to complete a short questionnaire regarding their activities and their perceptions of the class activities. The ESM procedure is particularly effective because it allows students themselves to report on what is occurring in the classroom periodically throughout the entire mathematics or science lesson. During classroom instruction, students were asked what they were doing and how they were feeling, while we observed the classroom activities that were taking place. The questionnaire was organized to measure the level of students' engagement, using eight survey items, including "I was paying attention" and "I was completely into class." In addition to other items related to students' perceptions and reports on classroom activities, two researchers sat in classes and took notes on what was occurring between signals.

Further, to supplement data, we conducted a focus group interview with ten randomly selected students from among ESM participants at each high school that we visited. After observing during a week of data collection, the group interview offered researchers an opportunity to directly ask students what they liked about their classrooms and what engaged them in learning. Interviews helped us understand the subjective world of students, and in particular, their understanding of what constitutes good teaching and learning.

Class Activities

Previous research and our classroom observations in this project led us to focus on three general types of organization for classroom instruction: whole-class instruction (lecturing), small-group work, and individualized work. During our field observations in classrooms, the predominant instructional practice was whole-group, teacher-centered direct instruction. Table 6.1 presents the relative frequency of each configuration (percentage of time). When considering the configuration in an average classroom, students were doing group work 13 percent of the time when signaled, seat work 35 percent of the time, and the remainder of the time were engaged in whole-class instruction or lecture (52 percent). Regression coefficients from a multilevel model in which the level of student engagement is the outcome variable and the class configurations are the predictors indicate that students were more engaged when signaled during small-group work relative to whole-class instruction (β = .18), but were slightly less engaged during individual seat work (β = −.01). These regression models suggested that group work is more engaging in mathematics classrooms (β = .28) than in science classrooms (β = .13), and that individual work is not very engaging in either content area. These data reveal differences in the typical classroom configurations observed in classes in the four cities we studied. For example, in El Paso individual seat work was the most frequent mode of instruction when students were signaled, yet El Paso students in our study were most engaged during less frequently occurring small-group work.

The experiences described by students in our focus group interviews support the positive effect of working in small groups on students' engagement. During group work, students were encouraged to use different approaches to learning, including more student-centered, problem-solving strategies. Many students reported gaining greater understanding by actually "trying out" various problem-solving activities with their peers rather than listening to their teachers. In response to a researcher's question about why science class was interesting, a female student in Chicago replied:

> Well, this week we were working on labs such as . . . we were trying to find out stuff that has acids and stuff that was tomatoes, eggs, and lemon juice . . . [Interviewer: And why was that really interesting for you?] . . . I wanted to know what, you know, what kind of . . . each thing was; I wanted to know what they have inside them. (CH310)

Students thought group work was especially valuable because it allowed them to work with their peers. This point is summarized by a student who said, "Well like sometimes it's [having friends is] good 'cuz you know, if you don't know, . . . something [they will help out]" (MI310). For an El Paso female student, working with friends provides motivation to do the work. She said,

> But if I'm me, for example, if I'm by myself doing my work, then I just think for myself, I don't feel like doing it no more. At least [when doing

Table 6.1.
Effects of Class Configuration on Student Engagement

Effect†	Whole Sample		Math		Science		Chicago		Memphis		El Paso		Miami-Dade	
No. Beeps	1709		800		909		429		441		543		296	
	%	β	%	β	%	β	%	β	%	β	%	β	%	β
Intercept		-1.1		-1.6		-1.2		-0.96		0.21		-0.03		-0.42
Small group	13	0.18**	6	0.28*	20	0.13+	9	0.15	31	0.03	0.3	0.53	12	0.35*
Individual work	35	-0.01	38	-0.03	31	0.03	17	0.05	37	0.05	51	-0.17*	27	-0.03

+ = p < .10; * = p < .05; ** = p < .01; *** = p < .001.
† Whole-class instruction is an omitted category in this regression model.

139

group work] you have a partner that's like, c'mon let's do this, finish it up and then we can talk and then I'll finish it a lot faster than all myself. (EP310)

However, in most focus group interviews students were quick to point out the possible risks when working with friends. A male tenth-grade mathematics student in Miami-Dade said, "If you don't know something you ask your friends. But then it also helps you get distracted from what you're doing" (MI310). Our focus group discussion in one of our Chicago high schools elaborated this point. A male student pointed out:

If you pick somebody and you know they're not going to do anything, you might as well not pick that person . . . Because there's certain categories in the classroom, the doers, the maybes, and the don't doers and you have to know. (CH310)

Group work increased students' engagement level in another way. Students across our sites asserted that group work is effective because it is interactive. When students have questions in the context of a small problem-solving group, everyone can function as a teacher. When the whole class is working as a unit with the teacher lecturing, lessons are much less interactive and do not readily provide opportunities for students to gain the desired level of understanding. As an example, during our observations a misalignment of teacher and student activities and understanding occurred when teachers were covering a new step in problem solving but students were still copying or working on previous steps the teacher had just covered using the chalkboard or overhead projector. Some teachers seemed not to notice student difficulty in processing information during times when they were both hearing about a new topic and monitoring previous work. An eleventh-grade student in El Paso expressed his frustration with a teacher "who talks and talks" or "who talks to himself": "Cuz if I got to know something I just write it and write it and write it and don't actually get to sit there and read through what I just wrote, then I won't understand" (EP310).

Students complained about teachers "who do not teach," creating a situation where exasperated students simply waited until the teacher finished lecturing. A female student in an El Paso classroom explained why students were quiet during teacher monologues. Students hoped the teacher would finish, allowing them the opportunity to finally ask questions. In another science class in Chicago students complained during the focus group interview as well as after and even during class that some teachers did not teach and just "talked to themselves." In one such class we observed students doing homework for other classes or chatting with classmates while the teacher lectured, switching to a listening mode only when the teacher provided answers to the seat work assignment they were given for the day.

In summary, students would say that teachers are incompetent if they are not able to explain themselves well. This state of affairs seems to stem from occasions in the classroom when a teacher is lecturing; thus, lecturing, as an instructional strategy, is more likely to be associated with teachers "talking to themselves." Although many students argued that they like group work, this does not necessarily lead to the conclusion that teachers should exclusively provide students with problems and applications to carry out in small-group settings. A closer examination of the information we gathered from students during our focus group interviews helps augment our understanding of this issue. As far as students are concerned, group work allows them to process information leading to a more comprehensive understanding of the curriculum. In addition, group work provides greater opportunity to interact, try out problems through hands-on activities, and have ample time to ask questions to gain clarification. However, regardless of class configuration types, students find satisfaction in understanding the material presented in the curriculum and tend to evaluate teachers in terms of how well they are able to help students comprehend it.

Most focus group conversations concerning teachers in all cities revolved around teachers that students regard as good at helping them understand the content of the ongoing lesson. Students were most enthusiastic about teachers who made themselves available during lunch periods or after school because of the support they provided in understanding material students were not able to grasp during class. Good teachers helped students in several ways. First, good teachers allowed time either before or after school to assist struggling students:

> But then you got some teachers knowing you don't know . . . how to do it real good, they'll come [and] take their time out their schedule, but they'll take their time and ask the student, since you got a problem with this you need to have more time and me helping you so I won't help you through class, slow you down. (ME330)

A female student in El Paso explained that her teacher received the best teacher award around the time we visited her school because:

> She helps us, she explains it, she gives us notes, she shows us how to do it step by step and she doesn't just tell us go do it by yourself. I mean,

> if we need help we can go to her and she is always there in the morning if we need help with something. (EP310)

Students also liked group work because it allowed them to achieve a higher level of understanding by exploring with hands-on experiences, interacting with peers, and finding out things they did not know before. We interpret what students shared with us during the focus group interviews to mean that regardless of how the teacher organizes the classroom, students appreciate any teacher who can help them gain a better understanding of the lesson at hand. This interpretation may sound obvious, yet it has implications for how we might understand situations we frequently observed in our classrooms. At least in students' subjective worlds as they were described in our focus group interviews, teachers were blamed for not engaging students and praised for bringing students to richer and deeper levels of understanding. This was a consistent, collectively shared view students offered to explain their inability or, conversely, their ability to remain consistently engaged.

The Impact of Block Scheduling

Time spent learning complex material in mathematics and science is critically important for students' deep understanding of these subjects. However, it is not just time by itself that is critical; it is how that time is spent. School districts vary in how they organize classroom time; one startling finding from our work was the overwhelmingly large proportion of time spent by students in completing individual seat work (for example, 51 percent in El Paso). According to our field notes from El Paso, this large amount of time spent completing worksheets and the like seemed to arise from the fact that El Paso schools used block scheduling to provide a class period of ninety minutes. In classes with block scheduling across all sites we saw in both mathematics and science classes lecturing or what we typically associated with "teacher monologues" twenty to thirty minutes in duration. Students were then given a class assignment or asked to start their homework. For another twenty to thirty minutes, students worked on that assignment. When they finished, students typically spent the remaining twenty to thirty minutes socializing with each other. The tacit understanding appeared to be that when students were finished with an assignment, their obligation to the class was over. During seat work, teachers occasionally helped students with questions, but they typically sat at their desks and appeared to be doing administrative work or correcting homework or tests. In other words, by assigning students seat work for a majority of class time in a ninety-minute class, teachers were able to do work they would otherwise have been required to do outside class hours.

Students asserted that ninety-minute class periods provided an easier pace than classrooms governed by the traditional fifty-minute period. One female student said, "And I like our way of learning [referring to block scheduling]. I wouldn't change it because . . ." as another female student interrupts with, "Working to the bell is not for me." The first student continues, "working to the bell means sometimes you won't have enough time to finish; the teacher won't have enough time to finish their lesson" (EP310).

Socialization time when individual students had completed their homework seemed to be the order of classroom events for these students. Responding to a researcher's comment that some students may complete their seat work assignments quickly simply to allow time for socializing, a female student explains:

> But if you really care about your grades you're gonna . . . it should be a motivation like . . . they're already done, let me [also] try to finish so I can start socializing. With our class, it's usually when the teacher gets done that you have time for work or hanging out or whatever you want to call it. (EP310)

Students in these El Paso classrooms were limited in the amount of time they actually did classwork compared to students in classrooms of fifty-minute duration. The large proportion of class time taken up by chatting and socializing in these ninety-minute classes scheduled twice in a week exposed students to forty minutes of instruction every week. In contrast, in classrooms where students attended a fifty-minute class every day, students were exposed to a minimum of one hundred minutes per week of classroom instruction. Thus, as these and other analyses show, block scheduling of ninety-minute classes did not insure that learning occurred throughout this time period.

Is this phenomenon characteristic of El Paso only? Using ESM data, we wished to confirm that limited classroom instructional time is a consequence of ninety-minute periods. If this were true, we should see the same trend in other classrooms employing block scheduling, as did some schools in Miami-Dade and Memphis. We divided our ESM sample into classrooms with ninety-minute class periods versus classes lasting fifty minutes. The pattern revealed by the sequentially collected student responses is clear. In ninety-minute classes, engagement drops by almost .20 SD after the fifth beep, or at a fifty-minute interval. In contrast, students' engagement in fifty-minute classes showed greater stability throughout the period. In other words, a block scheduled ninety-minute class was not effective in holding students' attention after about fifty minutes, corresponding to the traditional fifty-minute period. However, the traditional class period was never quite as engaging as the first half of a block period.

We also see differentiation in the frequency of various patterns of classroom organization observed among participating teachers. We present basic descriptive statistics twice, first for the whole sample and second for the sample after we removed one teacher, Charles Taylor (see chapter 5) from Memphis, because his approach to teaching was characterized by large amounts of time devoted to group work. This coupled with extremely high levels of student engagement made him an outlier.

One clear distinction between block and traditional class periods is the percent of time devoted to individual seat work. Block scheduled classes placed students in seat work activities almost half the time, while students in traditional (fifty-minute) classes participated in seat work less than one-third of the time. More time is devoted to teacher lecturing in fifty-minute class (53 percent versus 46 percent of time). In block scheduled classes there is more class time during which researchers reported the absence of any instructional activity, thereby limiting students' access to instructional content. In both traditional and block schedule classes students reported that lesson content was new knowledge about 40 percent of the time. Students work in small groups more frequently during traditional class periods, while students in block schedule classes report more frequently that what is being learned is easy rather than hard.

From our analyses we also observed that conversation among students is strongly affected by the length of the class period. Observation notes indicate that students in ninety-minute classes socialize more with their student peers during or after prolonged periods of seat work. In the traditional fifty-minute classes students spent considerable time talking with peers about classwork and they did so much more than their counterparts in block scheduled classes. Students in traditional classes also chatted with their peers much less frequently than those in block scheduled classes.

The differences between the two types of classes detailed here as a whole are not as dramatic as our observations in El Paso classrooms might have led us to believe; nonetheless, the trend is consistent—students in ninety-minute classes are exposed to less instructional time, spending more time chatting, doing very little group work, and spending almost half the class time engaged in individual seat work.

Opportunity to Learn

Next we describe students' perceptions of what was being taught in their mathematics and science classrooms during our ESM observations. While student focus group interviews were a good source of information, student views in this context tended to be very similar across time and location. Our ESM statistics, in contrast, capture dimensions of students' opportunity to learn during the classroom lesson in question. The concept of opportunity to learn refers to several aspects of the learning process, such as the amount of coverage that teachers undertake of the subject matter at hand. In their study of classroom learning, Barr and Dreeben (1983) found that, even in the same classroom, opportunity to learn can vary by student membership in a given track or instructional group. Students tracked in subject matter areas such as general science will be denied access to powerful concepts, strategies, and ideas provided to their peers in physics in the same high school. Researchers argue (and we concur) that tracking and course-taking patterns are important organizational processes that differentiate students' opportunity to learn. Students placed in advanced classes are more likely to be engaged in learning more complex and sophisticated mathematics and science, are more resourceful in their approach to learning, and have teachers who hold higher expectations for their achievement (Gamoran, 1987; Stevenson, Schiller, and Schneider, 1994).

Our ESM questionnaire included the concept of opportunity to learn defined as the degree to which students are introduced to new content during mathematics or science instruction. Theoretically, students could spend long hours in class without being taught anything new, while other students might be exposed to new instructional content and yet spend little time in the classroom. Indeed, previous research (Stigler and Hiebert, 1999) shows that, in comparison with Japanese and German classrooms, American mathematics and science classroom instruction at the eighth-grade level covers little novel material. Our ESM data allowed us to conclude that this claim also holds true in the sample of urban high schools in our study.

The ESM survey asked students whether what was being taught was something they already knew, or they did not already know. Results indicated that nearly half of the time students reported that "what they were being taught" covered content with which they were already familiar (46 percent). They reported that 41 percent of the time content covered something they didn't know and 13 percent of the time they could not tell if what was being taught was new or review. The magnitude of these percentages is difficult to evaluate as review of previously cov-ered material is often the foundation for new knowledge. Results indi-cated there was more review in mathematics than in science and as a result the percentage of new material was higher for science. Among districts, more novel content was reported in Chicago, while in both Memphis and El Paso review predominated. Not surprisingly, in Chicago, where students report a higher percentage of new material, levels of student engagement were highest. Engagement appeared to be highest when the gap between review and new content was widest. As men-tioned earlier, El Paso classrooms were particularly likely to review information because students reported that teachers presented previ-ously covered content a third of the time. This might stem from the fact that, as discussed earlier, El Paso teachers devoted more time to seat work and little time to small-group activities.

Students were exposed to new knowledge most when they were doing small-group work (57 percent of the time) or while being lec-tured (41 percent of the time). Students working individually on seat work were presented new information only 31 percent of the time. As our focus group interviews revealed, these outcomes may be related to the fact that during group work students were able to do more hands-on activities and discuss novel ideas because they were allowed to figure out things together with their peers. When examining the level of engagement and the relationship between class configuration and what was being taught, exposure to familiar content was more engag-ing (54 percent) than novel material (31 percent) during individual work. While students report being more engaged during whole-class instruction when new material was being taught (41 percent), they were most engaged when working in small groups on novel material (57 percent).

Relevance of Instructional Content

We explored students' perceptions of the relevance of instruction and changes in their engagement in terms of the content's relevance to students' everyday lives, plans for their future education, future jobs, and other factors. Overall, the most important category among items related to engagement was students' perceptions of the relevance of instructional content to tests. When students perceived that instruc-tional content was related to tests, they were engaged 87 percent of the time, but when instructional content was unrelated to tests, their engage-ment levels decreased by .13 SD. Chicago stood out in terms of this effect in that when students felt the class was unrelated to a future test, which occurred only 10 percent of the time, student engagement drastically decreased by .50 SD.

College was also seen as related to instruction 64 percent of the time. El Paso students reported .15 SD higher engagement level when instructional content was relevant to future college plans. Relevance to everyday life and jobs was also reported about a third the time. Everyday life, in addition, was positively related to engagement. Overall, students who judged that the class was relevant to their everyday lives reported .15 SD higher engagement levels. Miami-Dade and Memphis showed even higher levels of engagement (.23 SD and .25 SD, respectively) if students perceived relevance to their everyday lives. Class relevance to jobs had a significant effect of .32 SD in Miami-Dade. It is interesting to note that in Miami-Dade relevance to college attendance reduced the engagement level, while relevance to future jobs increased it. It is possible that Miami-Dade students were more aware of jobs than of going to college as immediate future options, which may have influenced how they experienced their mathematics and science course work.

The relevance of tests for student engagement was explored further by examining four types of tests: class quizzes, term tests, SATs, or state-level assessment tests. The results are somewhat startling; the largest impact on student engagement occurred when students were engaged in preparation for state-mandated, high-stakes tests. Although students believed that these tests were related to the content taught only 19 percent of the time, students reported higher levels of engagement when they thought class work was relevant to state-level exams.

Students complained during interviews that high-stakes state exams made them nervous, largely because they were given only a single opportunity to pass them. In Miami-Dade, a student said, " . . . but the one test that if you fail, you don't graduate or you get a thing [Certification of Attendance] of participation—that's messed up. You go over there for twelve years and all you get is a thing of attendance—that's messed up" (MI310). The high-stakes nature of the state-level exam is more than likely related to higher levels of students' concentration on what was being taught.

Students were uniformly aware of the importance of all tests. What follows is a conversation that occurred in a Miami-Dade high school about state-level high-stakes testing:

> Student A: [From test results] Um, you know how much you have learned, how much you have paid attention in your class. So it's a way to shake yourself, you're up to the level where you were expected to be.

Student B: I agree with him. 'Cuz tests are important 'cuz it's like to . . . testing to know how much you're learning through the years.
Student C: I also agree [with] what they both said because I, . . . what you learn there is like a [unclear] you know if you actually understood what you, what they told you to do. And if you don't understand you don't pass.
Student D: I think tests are a way of finding out whether or not you're ready to go out into the world and get a job or go on to higher education like college. (MI310)

While we are concerned with the possibility that state-level exams make learning processes test-driven, students may find it an opportunity to test their ability and learn how they stack up with others in their school, city, and state.

Class Configurations and Student Perceptions

Next we examine students' perceptions about how the content of their classroom experiences varied according to classroom organizational patterns determined by their teachers. Focus group interviews suggested that class configuration influences the way students experience classroom instruction, particularly with group work, an important vehicle that can allow students to try out novel ideas while interacting with their peers. We found that, particularly during small-group work, students were more likely to be exposed to new instructional information. When class organization and instruction's relevance to life are considered, students report similar levels of engagement across different instructional configurations. With regard to the relevance of classroom material to students' everyday lives, however, students report relevance 54 percent of the time in small-group activities compared to when they are working as a member of the whole class (33 percent) or in individual work configurations (39 percent). The effect of relevance to everyday life on engagement level remained positive and relatively large.

Professional Development and Classroom Practice

We evaluated the links between professional development reported by teachers and students' classroom experiences. Our participating teachers were involved in varying degrees with professional development provided by the district's USI program and designed to promote standards-based teaching and encourage supportive interactions between students and teachers. To the extent that professional development was successful with our participant teachers and promoted standards-based

Table 6.2.
Impact of Teacher Professional Development Activities on Student
Perceptions of Lesson Relevance

	Whole Sample		*Low PD Impact*		*High PD Impact*	
Effect	*Percent*	*Coefficient*	*Percent*	*Coefficient*	*Percent*	*Coefficient*
Intercept			—	−1.75	—	−0.57
Class Configuration						
Small group	13	0.18**	6	0.51***	21	0.09
Individual Seat work	35	−0.01	38	−0.10$^+$	30	0.08
Ratings of Knowledge						
Already knew	46	0.35***	48	0.35***	43	0.33***
Did not know	41	0.23***	40	0.28**	42	0.13
Uncertain	13	0.00	12	0.00	15	0.00
Ratings of Relevance						
Everyday Life	37	0.15**	33	0.14$^+$	43	0.12$^+$
College	64	0.01	63	0.08	65	−0.02
Future Job	31	0.06	31	0.19*	32	−0.06
Unrelated	13	−0.13*	13	−0.12	14	−0.17$^+$

$^+$ = $p < .10$; * = $p < .05$; ** = $p < .01$; *** = $p < .001$.

teaching, we should detect the differences made by the reform-oriented teachers in our ESM data.

We separated ESM participant teachers into two groups: those who stated that they had been influenced by professional development activities and those who stated they had not. The SCP asked teachers to rate the impact that they felt for twelve professional development activities. For example, teachers were asked to rate the impact of learning how to implement state or national content standards or participating in a study group on improving teaching. We combined these items and used an average as a composite of the impact of professional development activities. Using the median score as a threshold, teachers were grouped into high-impact and low-impact categories of professional development.

Table 6.2 reports the results of the separate regression models that predict student engagement for high-impact and low-impact teachers. Intercept values correspond to the predicted mean engagement of a student who said "no" to all the survey items used in this regression model. As we would expect, teachers reporting high impact of their professional development work have higher intercepts ($-.57$ versus -1.75), meaning that high-impact teachers foster greater levels of student engagement, when the other factors are controlled.

The results shown in Table 6.2 also suggest that teachers reporting a higher impact of professional development on their teaching practices use more small-group work (21 percent as opposed to 6 percent) and less seat work (30 percent as opposed to 38 percent) than teachers who report little or no impact. In terms of differential effects on student engagement levels, although low-impact teachers used small-group work only 6 percent of the time, they engaged students more when they did use group work ($\beta = .51$). Conversely, low-impact teachers used seat work 38 percent of the time, and when they did so students were less engaged ($\beta = -.10$).

As to whether teachers expose students to new knowledge, we did not see any large differences between those teachers reporting high levels of impact on their instruction from professional development experiences and those teachers reporting little to no impact. They did not differ notably with respect to student reports of the relevance of classroom lesson content; however, students who were instructed by teachers highly affected by their professional development experiences were more likely to see a connection with their experience outside the classroom, especially with reference to their everyday life (43 percent of the time versus 33 percent for low-impact teachers).

Looking at Individual Teachers

Up to now we have seen from the results of our regression models that teachers who see their professional development as having a greater impact also engage in more effective instructional activities. By using a multivariate model, we can see the phenomenon in abstract ways; however, this removes us from classroom realities. Instead, in those types of analyses researchers tend to talk about variables as if the variables were the major actors.

Table 6.3 summarizes important statistics for sixteen teachers from the four cities, whose students participated in the ESM study. Teachers were ranked by the mean engagement level of students and, for each teacher, statistics on key variables such as professional development impact, class configuration types, and novelty of the teaching content (whether students did not know the content). One immediate point to notice is that the differences among teachers were not so great. In fact, the use of the three-level Random Intercept Model with no Predictors tells us that in our sample the teacher-level variance is only about 10

Table 6.3.
Student Engagement in Classrooms with High or Low Professional Development Impact

						Percent					
Site*	Teacher**	Subject	Block	PD Impact†	Engagement Level β	Lecture	Small Group	Seat Work	New Knowledge	Easy Work	No Work
ME	Taylor	S	Yes	High	0.86	27.3	68.4	31.6	61.9	18.7	0.0
EP	Quinn	M	Yes	Low	0.45	27.5	0.0	77.8	16.8	8.1	0.0
ME	Upton	M	Yes	Low	0.39	44.2	3.9	72.1	37.0	33.1	3.2
CH	Evans	M	No	Low	0.34	50.0	0.0	36.6	34.8	8.0	16.1
ME	Smith	M	No	Low	0.27	57.3	23.6	25.4	46.4	9.1	5.4
MI	Bridges	S	No	High	0.26	15.6	8.9	75.6	20.0	24.4	0.0
ME	Holloway	S	No	High	0.07	15.1	48.8	33.1	35.5	17.5	3.0
CH	Solomon	S	No	Low	-0.03	86.9	21.5	03.8	60.8	14.6	0.0
EP	Nichols	S	Yes	Low	-0.18	37.2	0.0	55.6	35.8	11.1	9.2
MI	Specter	M	Yes	High	-0.19	91.1	17.8	7.8	35.6	22.2	0.0
CH	Thomas	S	No	High	-0.22	78.4	0.0	26.1	27.0	2.7	0.0
EP	Carter	M	Yes	High	-0.22	49.1	1.2	46.8	37.4	15.2	27.5
MI	Thornton	M	Yes	High	-0.34	71.3	0.0	28.7	47.5	9.0	0.0
EP	Sanchez	S	Yes	Low	-0.36	40.6	0.0	54.5	39.6	12.9	11.9
MI	Fuller	S	No	Low	-0.37	56.2	22.5	30.3	48.3	21.4	0.0
CH	Kurtz	M	No	Low	-0.46	53.0	26.0	31.0	40.0	5.0	9.0

* CH = Chicago; EP = El Paso; ME = Memphis; MI = Miami-Dade.
** Pseudonyms.
M = Mathematics; S = Science.
Block = Use of ninety-minute class periods.
†Low = Low reported impact of Professional Development; H = High reported impact of Professional Development.

percent, while the rest of the variance is shared almost equally by within-individual and between-individual differences.

Given that teacher differences are not as great as we might think, there is a trend, as confirmed also in an earlier regression presentation, that the higher the mean engagement of students, the lower the exposure of students to new instructional material. Some teachers appeared very engaging, yet they rarely introduced their students to new content. For example, her students found Taina Quinn (pseudonym) of El Paso highly engaging. She was described by students as really caring about them and as a teacher who made the learning process exciting by explaining mathematics problems step by step. However, she, like others teaching ninety-minute classes, lectured a quarter of the time and spent the rest of the period having students complete assigned seat work. During our classroom observations, students completed individual class work or homework on subject matter just covered in class while the teacher spent a majority of this time correcting assignments or doing administrative duties. In contrast, teachers whose data are arrayed at the bottom of the table with low mean engagement levels have statistical profiles that are not entirely negative. For example, all four teachers at the lowest end of this scale consistently presented new material to students as frequently as 40 percent of the time. In fact, some of these teachers' classroom activities appeared to not resonate well with their students given the low engagement levels.

Kathy Kurtz of Chicago is an example of a teacher using standards-based practices and without support from colleagues. During our week of observations she appeared to have a very engaging class, presenting a nice balance of small-group work, student presentations, and lectures. Her class started with an "opener" exercise, during which students solved a problem written on the board. Her class was very lively. She used a graphing calculator that projected the results on an overhead screen so students could see how mathematical functions appeared on the X- and Y-axes. She also had groups of students present solutions to problems that they had collectively solved. However, she also taught at a high school where she was one of only two or three teachers actually carrying forward standards-based instruction in a consistent manner.

CONCLUSION

What are the implications of our findings for classroom practices and policies of stakeholders in urban school districts that we studied? We believe there are a number of recommendations that would assist with improving classroom instruction and increasing students' opportunities to learn. One of our findings was that students attributed their levels of engagement primarily to their teachers' instructional performance. We may take what they are saying seriously and decide to devote more professional development time to techniques of teaching. Of course, other elements of the teaching profession, such as being concerned and passionate about students, are important. Students appeared to value such social and affective elements when these same caring teachers were able to help them improve their understanding of the content.

We found through ESM that student engagement levels were associated with various processes found in classroom practices. Group work was a particularly important factor. Students expressed the idea that group work encourages them to explore problems on their own, work together with peers, and provide them with different ways to understand the problems posed in mathematics or science classes, which all can be understood as ingredients for higher levels of engagement in classrooms. We also learned that group work introduces students to new sources of information, compared to other class configuration types. Students told us that group work allows them to work out problems on their own and find out things that they did not know before. Thus, based on these findings, professional development activities should stress the effective use of group work.

Our findings also tell us that perhaps there are limits to professional development activities targeted on content or particular pedagogical activities. We found through the use of ESM that teaching new instructional content occupies a smaller portion of class time than we expected. By this we mean that a majority of class time was devoted to reviewing material students said they already knew. Based upon our results, if we consider the amount of class time during which students have opportunities to learn new concepts and acquire new skills, we are left with the unhappy reality that current professional development activities are not translating into the classroom with effective instructional strategies and content.

While we would not argue that students must be exposed to new knowledge all the time how much repetition is most effective remains an unanswered question. Another complicating factor is that redundancy may occur not only within classes where teachers are reviewing material from previous lessons, but can also occur across grades and even across school levels. It would be helpful if future research could capture how much redundancy occurs as a result of a teacher's instructional style and how much occurs across grades or schools. If current policies do not foster a coherent national curriculum, as TIMSS researchers would have us believe, simply addressing the pedagogical approaches taken in a given classroom will not necessarily solve the problem of content redundancy in the classroom.

When we examined other factors that may to contribute to reducing students' opportunities to learn new material we noticed the importance of how classroom learning is organized. We found in some classrooms, particularly those that employed ninety minutes of instruction, that disproportionately more time was devoted to seat work. In these classes students routinely completed worksheets on material previously covered in class and then socialized after they finished these assignments, or the students would be expected to begin their homework to exhaust the time between the end of routine class activities and the end of the period. During these seat work periods, teachers would consult with the few students who sought help, but they typically would spent most of that time correcting homework or carrying out administrative tasks at their desks and tolerated varying levels of student socialization.

Looking at this problem from a broader perspective, we believe there may be unintended consequences of school- and district-level policies that influence classroom practices and students' subsequent learning. For example, schools that employed ninety-minute blocks of instructional time, where teachers spend proportionally longer periods on ineffective seat work activities, evidenced a loss of opportunity to learn new material. In our interviews with students we found that some students took it for granted that they could spend time talking to friends after completing their seat work during these ninety-minute lessons. Thus, policies that are not effectively implemented or monitored can make their way into the culture of classrooms, affecting how teachers teach and students learn, as well as how both teachers and students come to understand what constitutes classroom learning, thereby undermining the intent of professional development programs.

7

Closing the Achievement Gap

Arguably the most important aspect of the work we undertook over a three-year period in the four urban sites was to address the question of how well USI reforms in mathematics and science worked to close achievement gaps between groups of underserved students and their more privileged counterparts. Our aim throughout this research was to determine the impact of the NSF reform agenda on student achievement outcomes using NSF's six-driver model described in chapter 1. To answer this question we used a series of analyses combining qualitative measures such as classroom observations with teacher and student reports of instructional practice (see chapter 5) and measures of student achievement. Steps along the way included a path analysis for Miami-Dade and Chicago using a variety of data sources developed by reducing fifty-two different indicators into seventeen components to model systemic reform in urban schools. We surveyed teachers in forty-six schools to ascertain the level of professional community, a potential correlate of achievement (Louis, Marks, and Kruse, 1996).

From the Survey of the Enacted Curriculum (SEC) data, we identified five instructional factors (problem-solving activities, small-group work, hands-on activities, use of technology, and use of equipment), two assessment factors (performance and testing), four professional development factors (standards-based instruction tools, standards-based instructional methods, time in professional development, and preparation coursework), three teacher opinion factors (traditional beliefs, standards-based beliefs, and beliefs about sharing), three teacher preparation factors (equity, students with special needs, and use of standards-based practices), and two instructional influences factors (national and state standards, students and parents). Cronbach's alpha was calculated for each factor as a measure of reliability, and those factors considered in the next phase had alpha coefficients ranging from 0.59 to 0.95.

In this chapter we will describe quantitative analyses used in our three-year evaluation of NSF's Urban Systemic Initiative in the four cities, provide a discussion of path analyses of two models of systemic reform, and examine the results of these analyses as they address the question of what factors or indicators increase student achievement and close the achievement gap most effectively. Finally, a cross-site analysis provides indications of the extent to which our results may generalize to other operationalizations of systemic reform. These aspects of our work are critically important to educational policy makers at several levels in the system (Cohen and Ball, 1999; Cohen and Hill, 2000).

Causal Modeling

Path analysis was developed by Sewall Wright (1934) as a technique for disentangling direct and indirect effects of certain variables (hypothesized to be "causes") on other variables assumed to be dependent upon the values of causal variables. The technique is not intended to *discover* causal relationships, but rather to evaluate the tenability of a causal model that is identified a priori. The six-driver model proposed by NSF is one such a priori model that was evaluated using the data collected in the four USI sites (see Figure 7.1). An alternative model, incorporating a seventh driver, was also investigated. See chapter 8 for more information regarding the seventh driver, school culture.

Under a set of important assumptions (for example, an accurately specified model with linear, additive relationships among variables; fully recursive relationships; freedom from measurement error), the observed covariation between variables may be decomposed into direct, indirect, and spurious components by translating the causal model into a set of linear equations. If the assumptions are tenable, the parameters of these equations (path coefficients) represent the direction and magnitude of the causal connections between variables.

Figure 7.1. NSF Six-Driver Model of Urban Systemic
Reform for Miami-Dade

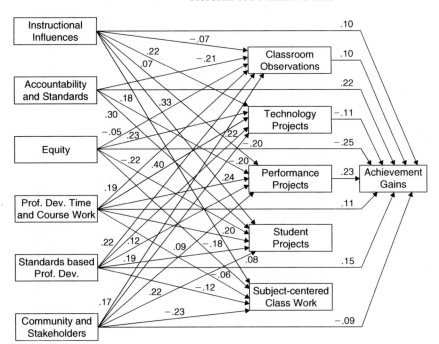

DEVELOPMENT OF THE MODEL

The variety of data sources incorporated in our study allowed us to develop multiple indicators for each of the NSF drivers (see Table 1.2 in chapter 1), to do so over a three-year period, and ultimately to execute a comprehensive analysis of multiple salient features fostering or inhibiting student attainment. In turn, these indicators informed our conceptualization of the NSF driver model, including a hypothesized seventh driver. Indicators included both organizational and individual level factors such as student achievement and student engagement (D5 and D6); teachers' reports of their professional development experiences, use of technology in the classroom, involvement in decision-making processes, and so on (D1 and D3); school district assessment practices and school-level support structures (D2); teachers' classroom practices (D1); the nature of community-school partnerships and other arrangements with business and industry (D4); school climate and school leadership (D3 and the hypothesized Driver 7 [D7]). Descriptive statistics, correlations, internal consistency, reliability, and path analysis informed our analyses. In addition, for the analyses reported here we computed student mathematics achievement gains over the period of the reform (1995–1999). The achievement gap was conceptualized in terms of both gender and race/ethnicity (Latino, black, and white). Our initial analyses described in this chapter are based on data gathered from students, teachers, administrators, and documents in the Miami-Dade initiative. Our models were then replicated using data from the Chicago initiative to investigate the extent to which causal mechanisms are common across reform sites.

MIAMI-DADE INITIATIVE

Data Reduction with Principal Components Analysis

The extensive sets of variables measured on students, teachers, and administrators were reduced using a two-step process. Initially, a series of principal component analyses were conducted to determine the structure of our instrumentation and to reduce our set of variables to a conceptually meaningful yet smaller subset. For example, the Study of the Enacted Curriculum: Survey of Classroom Practices in Mathematics and Science (SEC) (Blank, Porter, and Smithson, 2001) is a 155-item teacher survey with a companion measure for students (SEC) that has fifty-three overlapping items. Our principal component analyses yielded a set of five components to represent these items. Thus, each instrument was subjected to a principal components factor analysis and items with loadings greater than 0.40 were combined to create factor score estimates. A comprehensive list of components, the underlying factors, and the original items and estimates of factor reliability are presented in Table 7.1. This is an important table for the reader to consult since it provides working descriptions for the many components that were used in our analyses and anchors them to their respective measures such as the SEC, classroom observations, and the like.

In the second stage of the data reduction, the factors obtained from stage 1 were mapped onto the NSF drivers and a second set of principal components analyses was conducted. The component scores in this analysis were calculated using the factor score estimates. For example, D1 in the NSF model includes standards-based instruction.

Table 7.1.

Results of Principal Component Analysis for Miami-Dade Sample

Driver	Driver Name	Factor Name	Description of Factor Components	Source
Driver 1	**Standards-Based Instruction**			
Driver 1.1	Classroom Observation of SB Instruction	Modeling Connections and Reasoning $\alpha = .94^*$	Teacher observed modeling mathematics connections; guiding mathematics communication; modeling mathematics reasoning; and guiding students in making representations.	Standards-Based (SB) Checklist of Classroom Observations
		Connections $\alpha = .86$	Teachers observed guiding students with making connections and using visual connections in room.	SB Checklist of Classroom Observations
		Student Centered $\alpha = .58$	Teachers observed fostering student-centered communication and reasoning using subject-centered reasoning.	Analysis of Classroom Observations
		Problem Solving $\alpha = .90$	Teacher observed guiding students and modeling problem solving (PS).	SB Checklist of Classroom Observations
		[Homework]**	Teacher gives and counts homework and the type of homework activities.	Student Survey of Classroom Practices (SCP)
Driver 1.2	Technology Projects	Use Calculators $\alpha = .86$	Use calculators or computers as instructional activities; build models or charts.	Teacher Survey of Classroom Practices (SCP)

(continued)

159

Table 7.1.
(*continued*)

Driver	Driver Name	Factor Name	Description of Factor Components	Source
		Doing SB Activities $\alpha = .70$	Demo, presentation, or proof; use measuring tools; measure objects; analyze data for conclusions while PS; (solve word problems from text or worksheet).**	Teacher SCP Instructional Activities
		Worksheet	Complete problems from textbooks or worksheets; do arithmetic computations for homework.	Student SCP
		[Hands-on]*	Use hands-on materials or manipulatives such as counting blocks and algebraic tiles.	Student SCP
Driver 1.3	Performance Projects	Performance Items $\alpha = .85$	Student demonstrations; maintaining portfolios; working on projects.	Teacher SCP Assessment Practices
		SB Projects $\alpha = .91$	Collect or analyze data; solving problems; collect or analyze via Internet; projects outside the classroom; project lasting longer than a week; write a report; explain reasoning; use graphing calculators; (solve problems in groups).	Teacher SCP Instructional Activities
		Class Work $\alpha = .93$	Do arithmetic procedures; do computations from text or worksheet; show steps in problem solving; read about math (nontext); work with hands-on materials; work in groups; written assignment for text or work-sheet; use hands-on materials; write problem-solving explanations; (use computer tutorial software); (solve novel problems).	Teacher SCP Instructional Activities

(*continued*)

Table 7.1.
(continued)

Driver	Driver Name	Factor Name	Description of Factor Components	Source
Driver 1.4	Student Projects	Student Projects	Participate in projects outside the class-room; project lasting longer than a week.	Student SCP
		[Subject-Centered Representation] α = .42	Teacher observed using subject- and teacher-centered representation.	Analysis of Classroom Observation
Driver 1.5	Subject-Centered Classwork	Subject-Centered α = .60	Teacher observed using subject-centered communica-tion, problem solving, and connections.	Analysis of Classroom Observation
		[Classroom Activities] α = .61	Work individually in class; estimate, predict, or guess while PS; take tests; work in groups to improve work; watch teacher demonstrate PS; use manipulatives; collect data using hands-on materials; (presentation about concepts or projects); (take a test on a computer); (use portfolios); (apply PS to the real world).	Teacher SCP Instructional Activities
Driver 2	**Unified Policy**			
Driver 2.1	Instructional Influences	Students and Parents α = .59	Meeting the needs of students and parents and preparing for the next grade.	Teacher SCP Influence
		Standards α = .86	District, state, text, or national curricula; pre-service experience; district or state tests.	Teacher SCP Influence
		Goals and Mission α = .79	Quality of math and science goals; measurable assess-ments; mission statement; staff development plan.	School Improve-ment Plan
Driver 2.2	Accountability and Standards	Teacher Accountability α = .45	Impact of accountability; impact of SB instruction; (impact community resources).	Principal Interview

(continued)

Table 7.1.
(continued)

Driver	Driver Name	Factor Name	Description of Factor Components	Source
		Impact of USI and Standards α = .88	Impact of USI on science achievement; impact of standards on science instruction; (impact resource coordination).	Principal Interview
		Impact of USI PD α = .60	Principals' rating of the impact of math and science PD and USI impact on math achievement.	Principal Interview
		Technology α = .83	Principals' rating of USI impact on technology and the impact on achievement.	Principal Interview
Driver 3	**Unified Resources**			
Driver 3.1	Equity	Equity α = .90	Encourage minorities and females; teach, estimation, at assigned level, problem solving, diverse abilities, and varied cultural.	Teacher SCP Preparation
		Use S-B Practices α = .89	Meet standards; varied assessments; manipulatives; adapt curricula; integrate math with other subjects.	Teacher SCP Preparation
Driver 3.2	PD Time and Course work	Prof. Dev. Hours α = .85	Total time spent in PD in both content and pedagogy.	Teacher SCP PD
		Math Courses α = .86	Number of math and math education courses.	Teacher SCP PD
		Standards-Based Instruction α = .84	Technology; needs of students; journals; multiple assessment.	Teacher SCP PD
Driver 3.3	PD on SB Activities	SB Activities α = .77	Portfolios; networks; new methods; math content.	Teacher SCP PD
		Special Needs α = .81	Teach students with LEP, LD, or physical disabilities.	Teacher SCP Preparation

(continued)

162

Table 7.1.
(continued)

Driver	Driver Name	Factor Name	Description of Factor Components	Source
Driver 4	**Stakeholders**			
Driver 4.1	Community	Family α = .66	Parent and stakeholder involvement; mobility; and attendance based on family.	Evaluation of School Improvement Plans
		Community α = .66	SIP parents; community involvement; belief all children can learn; school demographics; (school partnerships).	Evaluation of School Improvement Plans
		Achievement α = .69	Increased achievement; advanced courses; plan to improve math/science.	Evaluation of School Improvement Plans
Driver 5	**Achievement**			
Driver 5.1	Math Scores	1996 Math Achievement	Math Achievement Scores.	Math96
		1997 Math Achievement	Math Achievement Scores.	Math97
		1998 Math Achievement	Math Achievement Scores.	Math95
		1996 Math Achievement	Math Achievement Scores.	Math96
		1998 Math Achievement	Math Achievement Scores.	Math98
Driver 5.2	Math Gain	1995–1997 Change 1995–1996 Change 1995–1999 Change	Change in Math Gain Scores. Change in Math Achievement Scores. Change in Math Achievement Scores.	MathDiff9597 MathDiff9596 MathDiff9599

(continued)

Table 7.1.
(continued)

Driver	Driver Name	Factor Name	Description of Factor Components	Source
Driver 6	**Achievement Gap**			
Driver 6.1	L-W-F	Latino Change	Change in Math Achievement Scores for Latinos.	DifHmaths95–99
		White Change	Change in Math Achievement Scores for Whites.	DifWmaths95–99
		Female Change	Change in Math Achievement Scores for Females.	DifFmaths95–99
Driver 6.2	M-B	Male Change	Change in Math Achievement Scores for Males.	DifMmaths95–99
		Change	Change in Math Achievement Scores for Blacks.	DifBmaths95–99
Driver 7	**School Culture**			
Driver 7.1	School Culture	Vision	Faculty and staff see themselves as having a shared vision.	I Culture Quality Survey (SCQS)
		Teamwork	Working together cooperatively in an atmosphere of mutual respect and caring.	SCQS[a]
		Facilitative Leadership	Faculty and staff view the administration of the school to be providing facilitative leadership.	SCQS
		Learning Community	Eagerly investing, learning, and working together toward their shared goals.	SCQS
		Culture of Sharing	Share; learn new things; make decisions; supported.	Teacher SCP Opinions
Driver 7.2	Teacher Opinions	Standards-Based Beliefs	Have curricular materials; all students and learn; peer planning; observe teachers.	Teacher SCP Opinions
		[Traditional Beliefs]*	Basic facts; basic skills; repeated practice.	Teacher SCP Opinions

*Cronbach's Alpha.

**Negative driver model factors and factor components are enclosed in parentheses.

164

The indicators for this driver were factors obtained in the first stage of analysis. These factors were derived from teacher responses to the SEC, student responses to the SEC, rubric scores from classroom observations, principal interviews, and school improvement plans. The second stage principal components analysis, summarized in Table 7.1, suggested five components to serve as indicators of D1 (classroom observations of standards-based instruction, technology projects, performance projects, student projects, and subject-centered class work). In total, the first stage of analysis yielded fifty-two factors, and the second stage reduced these to a total of seventeen indicator variables. The factor scores computed on these indicators were used in the subsequent bivariate correlational analyses and multivariate path analyses.

Correlations among Indicators

The correlations presented in Table 7.2 are helpful in clarifying conceptually the relationships among the indicators we used to measure the policy levers in the driver model. For example we report a negative relationship between classroom observation of standards-based instruction (D1.1 in Table 7.1) and teacher and student reports of instruction. This relationship suggests that researcher observations of instruction and teacher reports of their instructional practices regarding technology projects were somewhat at odds ($r = -.17$). The positive relationship between classroom observations and teachers' reports of standards-based professional development activities indicates that Miami-Dade teachers reporting participation in teacher networks and learning about portfolio use were observed modeling connections and fostering student-centered communication ($r = .21$). These practices as we know from analyses presented earlier in this volume are consistent with student academic achievement gains in mathematics.

In Table 7.2 the bivariate correlations among component scores suggest two areas for discussion. First, the component scores represent a set of independent measures of a conceptual group or *driver*. For example, standards-based instruction has five components with modest intercorrelations. In this group the strongest relationship is an inverse relationship between performance projects and subject-centered class work ($r = -.24$). Other relationships are slightly positive or slightly negative and are in the 0.1 range in magnitude. Two policy-related components (instructional influences and teacher accountability) have a near-zero relationship, suggesting that the two policy components are independent of each other ($r = -.06$). Relationships are modest yet positive among the unified resources grouping. Equity and professional

Table 7.2.
Correlation Matrix of Standardized Scores for Miami-Dade

Driver Component Name/Number	Driver Component Number																
	1	2	3	4	5	6	7	8	9	10	11	12	13	14	15	16	17
1. Classroom observations	1.00																
2. Technology projects	-0.17	1.00															
3. Performance projects	0.09	0.11	1.00														
4. Student projects	0.09	0.03	0.08	1.00													
5. Subject-centered class work	-0.10	0.12	-0.24	-0.08	1.00												
6. Instructional influences	-0.08	0.13	0.03	0.29	0.06	1.00											
7. Accountability and standards	-0.16	0.11	0.23	0.29	-0.09	-0.06	1.00										
8. Equity	-0.05	0.41	0.37	-0.07	-0.18	0.05	0.28	1.00									
9. Prof. develop. and course work	0.16	0.51	0.25	0.12	-0.07	-0.13	0.13	0.24	1.00								
10. Standards-based Prof. dev.	0.21	0.34	0.16	0.16	-0.15	-0.08	0.11	0.21	0.16	1.00							
11. Community stakeholders	0.07	-0.33	0.05	0.26	-0.17	0.13	0.11	-0.16	-0.25	-0.11	1.00						
12. Mathematics achievement	-0.13	0.55	0.15	-0.19	0.33	0.12	-0.29	0.29	0.19	0.12	-0.24	1.00					
13. Mathematics score gains	0.13	-0.02	0.23	0.15	-0.07	0.04	0.16	-0.06	0.11	0.14	-0.06	-0.17	1.00				
14. Achievement gap Latino, white, female	0.04	-0.09	-0.18	0.08	-0.03	0.01	-0.02	-0.18	-0.12	-0.13	0.77	0.04	-0.02	1.00			
15. Achievement gap male and black	0.09	-0.17	-0.12	0.09	-0.13	0.01	0.06	-0.26	-0.06	-0.18	0.68	-0.17	0.03	0.48	1.00		
16. School culture	-0.05	-0.25	-0.28	0.30	-0.08	0.00	0.13	-0.30	-0.15	0.05	0.00	-0.58	0.16	-0.07	-0.07	1.00	
17. Teacher opinions	0.05	0.36	0.15	-0.01	-0.03	-0.02	0.57	0.34	0.29	0.22	-0.05	-0.19	0.25	-0.04	-0.01	0.11	1.00

development course work are moderately correlated ($r = .24$), as are course work and instruction ($r = .21$). The strongest correlations are observed for the achievement gap related components ($r = .48$), while the remaining correlations are between -0.02 and 0.17. Correlations greater in magnitude than 0.35 are statistically significant at the .05 level.

Second, the correlations among component scores in different conceptual groups were somewhat mixed. The use of technology projects was positively correlated with each of the unified resources' indicators, including equity ($r = .41$), professional development course work ($r = .51$), and professional development related to standards-based instruction ($r = .34$). In addition, the correlations between each of the achievement gap indicators and the community stakeholder indicator were strong and positive in direction ($r = .68$ and .77).

Path Analysis

Path analysis expresses regression equations in the form of causal diagrams (as shown in Figure 7.1) to portray complex relationships among independent and mediating variables (D1–D4) that in turn explain outcome variables (D5 and D6). Such an analysis estimates the expected impact of the elements in the model on overall achievement and on the achievement gap in the context of systemic reform (Pedhazur, 1982). Path analysis allows the estimation of direct effects of causal factors, indirect (or mediated) effects, and total effects (the sum of direct and indirect effects). In the path analysis the contribution of the exogenous variables (such as policy, resource, and stakeholder drivers) on the endogenous variables (such as standards-based instruction and achievement outcome drivers) was determined. This analysis also determined the relative contribution of standards-based instructional practices as a mediating variable in the determination of student achievement.

Figure 7.1 presents the path model of student achievement gains based upon the six-driver model. In this graphical presentation the straight lines with arrows represent direct effects, while the curved double-headed arrows represent noncausal covariation (that is, the model asserts a correlation between indicators but does not hypothesize a causal direction). The path coefficients associated with each direct effect indicate the relative contribution (in both direction and magnitude) of each indicator in the causal chain. Indirect and total effects are computed from the direct effects illustrated. Because of the complexity of the graphical model, these effects are presented in Table 7.3 and are contrasted with the results obtained from fitting a seven-driver model (Table 7.4).

Table 7.3.

Six-Driver Model: Total Effects of Exogenous Variables (Policy, Resources, Stakeholders, and Instruction) on Endogenous Variables (Achievement and Instruction) for Miami-Dade

Endogenous Variables	Policy		Resources			Stakeholders	Instruction				
	Driver 2.1 Instructional Influences	Driver 2.2 Accountability and Standards	Driver 3.1 Equity	Driver 3.2 Prof. Dev. Time and Course Work	Driver 3.3 Standards-Based Prof. Dev.	Driver 4.1 Community Stakeholders	Driver 1.1 Classroom Observations	Driver 1.2 Technology Projects	Driver 1.3 Performance Projects	Driver 1.4 Student Projects	Driver 1.5 Subject-Centered Class Work
Instruction											
Classroom observations	−0.07	−0.21	−0.05	0.19	0.22	0.17	0.00	0.00	0.00	0.00	0.00
Technology projects	0.22	0.00	0.23	0.40	0.22	−0.20	0.00	0.00	0.00	0.00	0.00
Performance projects	0.07	0.18	0.00	0.24	0.12	0.09	0.00	0.00	0.00	0.00	0.00
Student projects	0.33	0.30	−0.22	0.20	0.19	0.22	0.00	0.00	0.00	0.00	0.00
Subject-centered class work	0.08	0.00	−0.18	−0.06	−0.12	−0.23	0.00	0.00	0.00	0.00	0.00
Achievement											
Mathematics achievement	0.13	−0.37	0.26	0.15	0.07	−0.13	−0.08	0.48	0.22	−0.05	0.32
Mathematics score gains	0.10	0.22	−0.25	0.11	0.15	−0.09	0.10	−0.11	0.23	0.00	0.00
Achievement gap											
Achievement gap Latino, white, female	−0.10	−0.14	0.04	0.08	−0.06	0.82	0.06	0.33	−0.20	−0.10	0.00
Achievement gap male and black	−0.03	0.01	−0.13	0.15	−0.11	0.69	0.11	0.19	−0.15	−0.15	−0.11

Exogenous Variables

Table 7.4.

Seven-Driver Model: Total Effects of Exogenous Variables (Policy, Resources, Stakeholders, Culture, and Instruction) on Endogenous Variables (Achievement and Instruction) for Miami-Dade

| | Exogenous Variables | | | | | | | | | | | | |
| | Policy | | Resources | | | Stakeholders | Culture | | Instruction | | | | |
Endogenous Variables	Driver 2.1 Instructional Influences	Driver 2.2 Accountability and Standards	Driver 3.1 Equity	Driver 3.2 Prof. Dev. Time and Course Work	Driver 3.3 Standards-Based Prof. Dev.	Driver 4.1 Community Stakeholders	Driver 7.1 School Culture	Driver 7.2 Teacher Opinions	Driver 1.1 Classroom Observations	Driver 1.2 Technology Projects	Driver 1.3 Performance Projects	Driver 1.4 Student Projects	Driver 1.5 Subject-Centered Class Work
Instruction													
Classroom observations	-0.08	-0.29	-0.07	0.17	0.21	0.17	0.00	0.15	0.00	0.00	0.00	0.00	0.00
Technology projects	0.22	-0.05	0.13	0.34	0.22	-0.22	-0.19	0.22	0.00	0.00	0.00	0.00	0.00
Performance projects	0.00	0.20	0.23	0.18	0.10	0.11	-0.20	-0.09	0.00	0.00	0.00	0.00	0.00
Student projects	0.33	0.41	-0.06	0.30	0.20	0.24	0.30	-0.37	0.00	0.00	0.00	0.00	0.00
Subject-centered class work	0.08	0.00	-0.27	-0.11	-0.11	-0.25	-0.18	0.13	0.00	0.00	0.00	0.00	0.00
Achievement													
Mathematics achievement	0.11	-0.17	0.16	0.12	0.13	-0.18	-0.48	-0.17	-0.07	0.46	0.11	-0.04	0.29
Mathematics score gains	0.07	0.09	-0.18	0.05	0.11	-0.07	0.08	0.21	0.06	-0.13	0.28	0.10	-0.02
Achievement gap													
Latino, white female	-0.08	-0.13	-0.08	0.05	-0.06	0.80	-0.06	0.15	0.07	0.33	-0.21	-0.12	0.00
Male and black	-0.03	-0.03	-0.20	0.14	-0.11	0.68	-0.03	0.11	0.11	0.19	-0.15	-0.15	-0.11

Six-Driver Model

The NSF six-driver model (as shown in Table 7.3) explained the association among aspects of standards-based instruction (classroom instruction, technology projects, performance projects, student projects, and class work), policy alignment (student, parent and principal views on professional development), and resources related to professional development. The six-driver model explained 16.6 percent of the variability in the mathematics achievement component. This explained variance was primarily attributed to three of the component groups—performance projects, technology projects, and standards-based practice observed in sample classrooms. Specifically, a one-standard deviation increase in use of performance projects is expected to result in a .23 standard deviation increase in student achievement gains over the period of the reform. Similarly, a one-standard deviation increase in use of technology projects is expected to result in a .11 standard deviation *decrease* in achievement, suggesting that students of teachers who use technology in their classrooms had smaller achievement gains than those who made less use of technology. Finally, a one-standard deviation increase in observations of standards-based practices is expected to result in a .10 deviation increase in student achievement. In general, the process drivers had small indirect effects on student mathematics achievement gains when direct effects and total effects were examined. In contrast, substantial direct effects were found for the policy component related to principals' view of the impact of professional development (.22), equity (−.25), time in professional development (.11), and professional development involving standards-based activities (.15). The total effect of instructional influences on overall achievement gains (summing the total effects across the five indicators of Driver 1) was positive (.22), while the total effect of stakeholders/community on achievement was negative (−.09).

We also modeled through path analysis an achievement model in which the components' content was largely comprised of gains for Latino, white, and female students (see Figure 7.1). This model, which explained 72.8 percent of the variability in student achievement, may be contrasted with a model whose components were related to score gains for black and male students ($R^2 = .58$). This suggests different explanations for reducing achievement gaps in student groups. This model (using success in reduction of the achievement gap as the outcome measure) suggested much stronger effects of certain driver components. For example, a one-standard deviation increase in the use of technology projects is expected to result in a .33 standard deviation increase in the achievement associated with Latino, white, and female students, but only a .19 increase in the achievement of African American students and males (Table 7.3). Similarly, a one-standard deviation increase in performance projects leads to a .20 standard deviation decrease in the Latino/white/female achievement component, and a .15 decrease in the African American/male component. The strongest effects are those associated with community stakeholders (.82 for Latino/white/female achievement and .69 African American/male achievement). These figures suggest the importance of stakeholder involvement for these groups and clearly underscore the importance of policy to encourage such involvement.

Seven-Driver Model

The total effects estimated from a comparable model, hypothesizing a seventh driver (school culture), are presented in Table 7.4. This model explained 19.9 percent of the variance associated with gains in student achievement. Four instructional components of this model had path coefficients larger than .05, technology projects ($-.13$), performance projects (.28), student projects (.09), and classroom observations of standards-based instruction (.06). These direct effects suggest that the use of performance projects has the largest positive impact on student achievement (in other words, a one-standard deviation increase in the use of performance projects in the classroom is expected to result in a .28 standard deviation increase in achievement gains). The direct effect of the instructional component that we term technology projects (perhaps a misnomer) is negative, indicating only moderate gains in students' mathematics achievement, and is associated with such traditional practices as using calculators and worksheets. These findings are in contrast to results associated with performance projects, including student demonstrations, student portfolios, use of the Internet, and opportunities to collect and analyze data—all activities and experiences that are rooted in standards-based, hands-on practices. Teachers who reported using calculators and worksheets did *not* report using the hand-on instructional activities associated with performance projects, and their students concurred with their teachers' self-assessments.

The results associated with the school culture component (the seventh driver) suggest substantial positive associations between a strong and supportive school culture and mathematics achievement gains, as we will discuss in more depth in chapter 8. Specifically, one-standard deviation increase in teachers' perceptions about a culture of sharing and beliefs about standards is expected to result in a .21 standard deviation increase in student achievement gains. Finally, a standard deviation increase in teachers' perceptions about the vision, team-work, facilitative leadership, and learning community of a school is expected to result in a .08 standard deviation increase in achievement gains.

The analysis of the achievement gap driver, using components comprised of gains for Latino, white, and female students and gains for black and male students provided total effect estimates for Drivers 1–4 (D1–4) that were consistent with those obtained using the six-driver model (Table 7.4). For example, in the seven-driver model a one-standard deviation increase in the use of technology projects is expected to result in a .33 standard deviation increase in the achievement associated with Latino, white, and female students, but only a .19 increase in the achievement of African American students and males. Identical effect estimates were obtained with the six-driver model. However, the school culture driver contributed to the explanation of gap reduction. Specifically, a one-standard deviation increase in teachers' perceptions about the vision, teamwork, facilitative leadership, and learning community of a school is expected to result in a .15 standard deviation increase in achievement gains indexed by the Latino/white/female component and a .11 standard deviation increase in achievement gains indexed by the African American/male component.

CHICAGO INITIATIVE

Data Reduction with Principal Components Analysis

The results of the components analysis conducted on measures obtained from the Chicago site are presented in Table 7.5. As with the Miami-Dade initiative, each instrument was subjected to a principal components factor analysis. Items with loadings greater than 0.40 were combined to create factor score estimates, and the factors obtained from stage 1 were mapped onto the NSF drivers followed by a second set of principal components analyses. The second-stage principal components analysis suggested five components to serve as indicators of D1 (instructional practices, solving problems and technology use, observed problem-solving practices, observed student-centered practices, and classroom instruction). Two indicators were evident for D2 (principals' views of reform impact and influences on policy), two for D3 (teacher preparation and professional development), and two for D4 (school mission and assessment plan). The outcome drivers (5 and 6) each evidenced two components reflecting gains from 1995–1999 and gains from 1995–2000. Finally, D7 suggested two components (school culture and

Table 7.5.

Results of Principal Components Analysis for Chicago Sample

Driver	Driver name	Factor Name	Factor Component Description	Source
Driver 1	**Standards-based Instruction**			
Driver 1.1	Instructional Practices	Small-Group Activities $\alpha = .92^*$	Talk about solving problems; written assignments from textbook or worksheet; assignment taking longer than one week; work with group to improve written work; review for a test or quiz (54–58).	Teacher responses to SCP
		Performance Assessment $\alpha = .88$	Use individual or group demonstration, presentation; use portfolios; use projects; use performance tasks or events (e.g., hands-on activities) (81, 83, 82, 80).	
		Traditional Assessment $\alpha = .70$	Use short-answer assessments; use extended response item for which student must explain or justify answer; use objective assessment items (78, 79, 77).	
		Technology Use Activities $\alpha = .57$	Learn facts or procedures; use sensors or probes; collect data or info using the Internet; display and analyze information; develop geometric concepts; take a test or a quiz; use individualized instruction or tutorial software (66–72).	
		Instructional Activities $\alpha = .86$	Watch a teacher demonstration; read about math (non-textbook); collect or analyze data; maintain and reflect on portfolio; use hands-on materials or manipulatives; engage in problem solving; take notes; work in pairs or small groups; activity outside the classroom; use computers or calculators; work individually; take a quiz or test (34–45).	

(continued)

173

Table 7.5.
(*continued*)

Driver	Driver name	Factor Name	Factor Component Description	Source
		Homework Activities $\alpha = .85$	Arithmetic computation or procedures; show required steps; explain reasoning or thinking; work on demo presentation or proof; collect data or info; write a math report (27–32).	Student responses to SCP
		Hands-on Activity $\alpha = .86$	Work with hands-on materials to understand concepts; measure objects; build models or charts; collect data by counting, observing, or surveys; present math information to students (60–64).	
		Problem-Solving Activities $\alpha = .73$	Computations from a text or worksheet; solve word problems from a text or worksheet; solve novel problems; write an explanation to a problem; apply math to real-world problems; make estimates, predictions, guesses, or hypotheses; analyze data to infer or conclude (23–29).	
		Small-Group Activity $\alpha = .49$	Talk about solving problems; written assignments from textbook or worksheet; assignment taking longer than one week; work with group to improve written work; review for a test or quiz (31–35).	
		Hands-on Activity $\alpha = .79$	Work with hands-on materials to understand concepts; measure objects; build models or charts; collect data by counting, observing, or surveys; present math information to students (37–41).	

(*continued*)

Table 7.5.
(continued)

Driver	Driver name	Factor Name	Factor Component Description	Source
Driver 1.2	Solving Problems and Technology	Problem-Solving Activities $\alpha = .54$	Computations from a text or worksheet; solve word problems from a text or worksheet; solve novel problems; write an explanation to a problem; apply math to real-world problems; make estimates, predictions, guesses, or hypotheses; analyze data to infer or conclude (46–52).	Teacher SCP Use
		Frequency of Equipment Use (Students) $\alpha = -.30$	Use manipulatives; use measuring tools; use calculators; Use graphing calculators (50–53).	Student SCP
		Frequency of Equipment Use (Teachers) $\alpha = .24$	Use manipulatives; use measuring tools; use calculators; use graphing calculators (73–76).	Teacher SCP
		Technology Use Activities $\alpha = .67$	Learn facts or procedures; use sensors or probes; collect data or info using the Internet; display and analyze information; develop geometric concepts; take a test or a quiz; use individualized instruction or tutorial software (43–49).	Student SCP
Driver 1.3	Observed Problem-Solving Practices	Guidance with Reasoning $\alpha = .87$	Teacher provides guidance to students about reasoning, communication, and representation of mathematics subject content.	Observation Checklist
		Student-Centered Communication $\alpha = .76$	Classroom interactions are more student-centered connections, communications, and problem solving.	Observation

(continued)

175

Table 7.5.
(*continued*)

Driver	Driver name	Factor Name	Factor Component Description	Source
		Problem Solving $\alpha = .81$	Students have opportunities to solve problems that go beyond following procedures and the teacher provides guidance and models problem solving.	Observation Checklist
		Subject-Centered Problem Solving $\alpha = .50$	Classroom interactions are more subject-centered connections, representations, and problem solving combined with teacher-centered problem solving.	Observation
Driver 1.4	Observed Student-Centered Practices	Student-Centered Connections $\alpha = .89$	Students have opportunities to make connections of subject content with other important topics, which are also modeled by the teacher.	Observation checklist
		Student-Centered Reasoning $\alpha = .74$	Classroom interactions are more student-centered representations, and reasoning combined with subject-centered communication.	Observation
				Observation
Driver 1.5	Classroom Instruction	Student Views of Classroom Instructional Practices $\alpha = .69$	Watch a teacher demonstration; read about math (non-textbook); collect or analyze data; maintain and reflect on portfolio; use hands-on materials or manipulatives; engage in problem solving; take notes; work in pairs or small groups; activity outside the classroom; use computers or calculators; work individually; take a quiz or test (11–22).	Student SCP

(*continued*)

176

Table 7.5.
(*continued*)

Driver	Driver name	Factor Name	Factor Component Description	Source
Driver 2	**Unified Policy**			
Driver 2.1	Principals' Views of Reform Impact	Standards-based Instruction $\alpha = .69$	Principal's definition of standards-based mathematics and science instruction.	Principal Interviews
		Changes Caused by USI $\alpha = .93$	Principals' views on changes resulting from USI implementation on professional development, technology, and community resources.	
		USI Benefits to Students $\alpha = .95$	Principals' views on how students have benefited from USI, including achievement.	
		Coordination of Resources $\alpha = .93$	Principals' descriptions of how mathematics and science resources are coordinated, including staff meetings.	
		School Plans for Achievement $\alpha = .64$	Clear statement of beliefs, plan for improving mathematics and science achievement, dealing with student mobility, involving parents, and staff development program.	School Improvement Plans

(*continued*)

Table 7.5.
(continued)

Driver	Driver name	Factor Name	Factor Component Description	Source
Driver 2.2	Influence of Policy	Impact of USI $\alpha = .99$	Principal's opinion of the impact of USI on math and science achievement in their school.	Principal Interviews
		Students and Community $\alpha = .85$	Influenced by students' special needs; influenced by parents/community; prepare students for next grade or level (92, 93, 94).	Teacher SCP
		Standards and Tests $\alpha = .45$	Influenced by district's curriculum framework or guidelines; influenced by state's curriculum framework or content standards; influenced by experience in pre-service preparation; influenced by textbook/instructional materials; influenced by national science education standards; influenced by district test; influenced by state test (86, 85, 91, 87, 90, 89, 88).	
Driver 3	**Unified Application of Resources**			
Driver 3.1	Teacher Preparation	Teach Mathematics $\alpha = .62$	Prepared to implement instruction that meets standards; prepared to use a variety of assessment strategies; prepared to teach the use of manipulative materials; prepared to select and/or adapt instructional materials to implement curriculum; prepared to integrate mathematics with other subjects; prepared to help students document and evaluate their own mathematics work; prepared to use/manage cooperative learning groups; prepared to involve parents in the mathematics education of their children (98, 99, 112, 111, 97, 101, 96, 108).	Teacher SCP

(continued)

Table 7.5.
(continued)

Driver	Driver name	Factor Name	Factor Component Description	Source
		Teach LEP α = .73	Prepared to teach students who have limited English proficiency; prepared to teach students who have a learning disability which impacts learning; prepared to teach students with physical disabilities (104, 105, 100).	
		Teach Diverse Abilities α = .84	Prepared to encourage participation of minorities in mathematics; prepared to encourage participation of females in mathematics; prepared to teach estimation strategies; prepared to teach at assigned level; prepared to teach problem-solving strategies; prepared to teach classes for students with diverse abilities; prepared to teach mathematics to students from a variety of cultural backgrounds (107, 106, 109, 95, 110, 102, 103).	
		Professional Development (PD) Activities α = .77	Participated in a formal portfolio assessment activity beyond your own classroom; participated in a teacher network or study group on improving teaching; new methods of teaching; in-depth study of content (141, 140, 135, 136).	
Driver 3.2	Professional Development	Course Work α = .77	Advanced mathematics courses (e.g., calculus, statistics); mathematics education; refresher mathematics courses (e.g., algebra, geometry) (146, 147, 145).	Teacher SCP
		Content and Pedagogy PD α = .85	Focused on methods of teaching mathematics; provided in-depth study of mathematics content; attended an extended institute or professional development program for teachers (cumulative 40 contact hours or more) (132, 131, 142).	

(continued)

179

Table 7.5.
(continued)

Driver	Driver name	Factor Name	Factor Component Description	Source
		Reform Agenda PD α = .62	Educational technology; meeting the needs of all students; read or contributed to professional journals; multiple strategies for student assessment; how to implement new curriculum or instructional materials; observed other teachers teaching in your school, district, or another district (139, 137, 144, 138, 134, 143).	
Driver 4	**Mobilization of Stakeholders**			
Driver 4.1	School Mission	School Mission α = .60	Mission statement and plan that accounts for school demographics and stakeholder involvement.	School Improvement Plans
		Achievement Measures α = .52	Clear plans for measurement of achievement and school attendance plans.	
Driver 4.2	Assessment Plan	Alternative Assessments α = .18	Clear plans that include alternative assessments to achieve mathematics goals and school partners.	
Driver 5	**Mathematics Achievement**			
Driver 5.1		Achievement Gains 95–99	Change in Mathematics Achievement Scores.	
Driver 5.2		Achievement Gains 95–2000	Change in Mathematics Achievement Scores.	

(continued)

180

Table 7.5.
(continued)

Driver	Driver name	Factor Name	Factor Component Description	Source
Driver 6	**Achievement Gap**			
Driver 6.1		Male/Female Gap 95–99	Change in Mathematics Achievement Scores for Males and Females.	
Driver 6.2		Male/Female Gap 95–2000	Change in Mathematics Achievement Scores for males and females.	
Driver 7	**School Culture and Opinions**			
Driver 7.1	School Culture	Teamwork	Working together cooperatively in an atmosphere of mutual respect and caring (1, 2, 4, 22, 23, 17, 11, 29, 8).	School Culture Quality Survey (SCQS)
		Shared Vision	Staff see themselves as having a shared vision (3, 5, 12, 21, 26, 35, 36).	
		Learning Community	Investing, learning, and working together toward shared goals (32, 31, 18, 34, 30, 27, 33, 28, 25, 15, 9).	
		Facilitative Leadership	Staff views the administration of the school to be providing facilitative leadership (10, 24, 14, 20, 19, 13, 7, 16, 6).	
Driver 7.2	Teacher Opinions	Culture of Sharing α = .04	Teachers regularly share ideas and materials; have many opportunities to learn new things about teaching; teachers contribute actively to decisions; are supported by colleagues to try out new ideas in teaching; receive little support from the school administration for teaching (122, 124, 126, 120, 121).	Teacher SCP

(continued)

181

Table 7.5.
(continued)

Driver	Driver name	Factor Name	Factor Component Description	Source
		Standards-based Beliefs α = .58	Have adequate curriculum materials available for instruction; all students can learn challenging content; have adequate time during the regular school week to work with peers on curriculum or instruction; teachers in this school regularly observe each other teaching classes; really enjoy teaching; Students learn best when they ask a lot of questions; required to follow rules at this school that conflict with best professional judgment about teaching and learning mathematics; absenteeism is a problem (128, 116, 127, 123, 119, 113, 125, 129).	
		Traditional Beliefs α = .43	Calculator should be used only after the mastery of basic arithmetic facts; it is important for students to learn basic skills before solving problems; students master and retain algorithms more efficiently through repeated practice than through the use of applications and simulations; students learn mathematics best in classes with students of similar abilities; mobility of students in and out of our school is a problem (115, 118, 114, 117, 130).	

* Cronbach's Alpha.
** Negative driver factors are enclosed in parentheses.

teacher opinions). In total, the data reduction resulted in seventeen indicator variables for the seven drivers to be included in the model.

There are several similarities and differences evident among the indicators extracted for Chicago and those obtained for Miami-Dade for each of the seven drivers. D1 for both sites described instructional practices and included a mix of more standards-based practices and traditional practices. Student views of homework practices countered those of their teachers in both sites. In addition, students described their class work as composed of "watching their teacher" and "doing arithmetic procedures." Chicago teachers and students were more likely to mention the use of technology and equipment, while Miami-Dade teachers mentioned more long-term project-based activities. Components involving researcher observations of instruction included more subject-centered activities in Miami-Dade, while Chicago components included more student-centered activities. Policy issues (D2) were derived mainly from School Improvement Plans and principal interview analyses.

Although analyses in both sites included factors regarding the impact of the USI on achievement and the influence of curriculum in mathematics and science and tests on instruction, Chicago components included the principal's role in the reform, while Miami-Dade's components were more focused on teacher and student accountability. D3, unified application of resources, included components describing teaching students with diverse abilities and teachers' professional development experiences for both sites. Chicago's components included more descriptions of pedagogy, professional development, and in-depth study of mathematics content. D4, mobilization of stakeholders, included plans for increasing student achievement at both sites, with supportive components of improving alternative assessments in Chicago. Miami-Dade's components focused more on increasing parent and community involvement. For our seventh driver, school culture, both sites had similar components. Analyses in both cases emphasized the importance of school teamwork and vision along with a school culture of sharing. In summary, both analyses identified the presence of important components of the reform agenda but also showed the prevalence of a number of lingering dubious constructs about student cognition and instruction. For example, a number of teachers at both sites still cling to the notion that students first must learn basic skills before solving complex problems.

Correlations among Indicators

Bivariate correlations among component scores (Table 7.6) suggest moderate to strong correlations among the five D1 indicators (ranging from $r = -.70$ between Observed Problem-Solving Practices and Classroom Instruction to $r = .68$ between Instructional Practices and Solving Problems/Technology Use). Similarly, strong correlations were evident between the two indicators of D2 ($r = .45$), the two indicators of D3 ($r = .84$), and the two achievement gap indicators of D6 ($r = .45$). In contrast, near zero correlations were obtained between the two indicators of D4, D5, and D7 (with correlations ranging from $-.07$ to $.06$).

Across the drivers, moderate correlations with achievement gains (D5) were evident for indicators in D2 ($r = .29$ and $.35$ for Principals' View of Reform Impact) and D3 (with rs ranging from $.21$ to $.39$). Moderate inverse correlations were observed for Solving Problems/Technology Use of the standards-based instruction driver ($r = -.29$ and $-.38$). A different pattern of bivariate relationships was observed for the achievement gap indicators (D6). Direct relationships were observed with Classroom Instruction ($r = .25$ and $.38$) and with Instructional Practices ($r = .28$ with 6.1, but nearly 0 with 6.2). Conversely, inverse relationships were observed with Observed Problem-Solving Practices ($r = -.27$ and $-.31$) and Teacher Preparation ($r = -.29$).

As was the case in our presentation of Miami-Dade analyses (page 165) we can also examine the bivariate relationships among the indicators to guide our understanding of reform in Chicago (Table 7.6). The correlations shown in Table 7.6 suggest that researcher observations of classroom instruction are different than teachers' reports of their own classroom instruction ($r = -.60$). However the pattern of relationships for the Chicago model indicators also reveals that students' reports of teacher instructional practices are quite similar to our own classroom observations ($r = .58$). Additional review of the relationships presented in Table 7.6 provides evidence of a strong relationship between standards-based professional development activities (a composite variable) and teacher reports of their preparation to teach diverse learners ($r = .84$) as well as positive relationships with gains in the level of mathematics achievement. Recall that standards-based professional development refers to participation in activities designed to equip teachers with skills in integrating mathematics with other subjects and in organizing students' cooperative learning as well as involving parents in supporting student learning.

Path Analysis

Figure 7.2 presents the path model of student achievement gains based upon the six-driver model estimated from the Chicago results. As

Table 7.6.
Correlation Matrix of Standardized Scores for Chicago

Driver Component Name and Number								Driver Component Number									
	1	2	3	4	5	6	7	8	9	10	11	12	13	14	15	16	17
1. Instructional Practice	1.00																
2. Prob. solving/tech	0.68	1.00															
3. Prob. solving practices	-0.60	-0.38	1.00														
4. Student-Centered Practices	-0.31	-0.27	0.44	1.00													
5. Classroom instruction	0.58	0.33	-0.70	-0.45	1.00												
6. Views of reform impact	0.11	0.04	-0.21	-0.47	0.24	1.00											
7. Influence on policy	-0.02	0.01	-0.20	-0.38	0.21	0.45	1.00										
8. Teacher preparation	0.05	0.17	-0.19	-0.26	0.18	0.18	0.86	1.00									
9. Professional development	0.03	0.03	-0.21	-0.33	0.28	0.33	0.93	0.84	1.00								
10. School mission	-0.02	0.05	0.18	0.02	0.08	-0.21	-0.18	-0.24	-0.05	1.00							
11. Assessment plan	0.13	-0.02	-0.08	-0.13	0.15	0.32	0.10	0.11	0.05	-0.06	1.00						
12. Achievement 1995-1999	-0.10	-0.29	0.06	-0.12	-0.15	0.13	0.29	0.21	0.27	-0.18	0.14	1.00					
13. Achievement 1995-2000	-0.22	-0.38	0.02	-0.00	-0.05	0.08	0.35	0.22	0.39	0.01	-0.14	-0.07	1.00				
14. Gender gap 1995-1999	0.28	0.08	-0.26	0.06	0.25	0.06	-0.20	-0.29	-0.19	-0.07	0.13	0.54	-0.48	1.00			
15. Gender gap 1995-2000	0.00	-0.07	-0.31	-0.17	0.38	0.10	0.01	-0.06	0.04	0.04	-0.15	0.18	-0.25	0.45	1.00		
16. School culture	0.04	0.04	-0.08	-0.01	0.00	-0.07	-0.06	-0.15	-0.11	0.08	0.08	0.11	-0.08	0.17	0.09	1.00	
17. Teacher opinions	0.94	0.65	-0.56	-0.43	0.62	0.16	-0.04	-0.01	0.03	0.05	0.14	-0.14	-0.19	0.25	0.01	0.06	1.00

185

Figure 7.2. NSF Six-Driver Model for Urban Systemic Reform
for Chicago

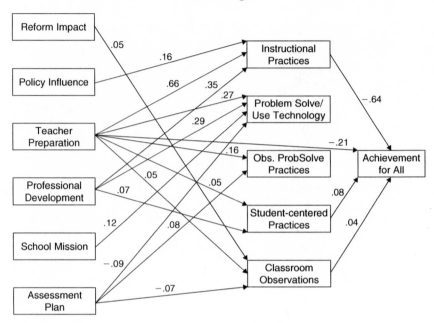

with the Miami-Dade results, the complexity of the graphical model makes tabular presentation more effective. The estimates of effects for the six-driver model are presented in Table 7.6 and are contrasted with the results obtained from fitting the seven-driver model (Table 7.7).

Six-Driver Model

The six-driver model explained 20.6 percent of the variability in the mathematics achievement component in Chicago and 26.8 percent of the variability in achievement gap. The explained variance in overall achievement was primarily attributed to three of the component groups—influence on policy, teacher preparation, and instructional practices. Specifically, a one-standard deviation increase in influence on policy is expected to result in a .09 standard deviation *decrease* in student achievement gains between 1995 and 1999, and .15 standard deviation *decrease* in achievement gains between 1996 and 2000. Similarly a one-standard deviation increase in teacher preparation is expected to

Table 7.7.

Six-Driver Model: Total Effects of Exogenous Variables (Policy, Resources, Stakeholders, and Instruction) on Endogenous Variables (Achievement and Instruction) for Chicago

| | Exogenous Variables | | | | | | | | | | |
| | Policy | | Resources | | Stakeholders | | Instruction | | | | |
Endogenous Variables	Driver 2.1 Views of Reform Impact	Driver 2.2 Influence on Policy	Driver 3.1 Teacher Preparation	Driver 3.2 Professional Development	Driver 4.1 School Mission	Driver 4.2 Assessment Plan	Driver 1.1 Instructional Practice	Driver 1.2 Prob. Solving/ Technology	Driver 1.3 Prob. Solving Practices	Driver 1.4 Student- Centered Practices	Driver 1.5 Classroom Instruction
Instruction											
Instructional practice	0.040	0.156	0.661	0.348	−0.011	−0.018	0.000	0.000	0.000	0.000	0.000
Prob. solving/tech	−0.084	−0.228	0.269	0.291	0.125	−0.087	0.000	0.000	0.000	0.000	0.000
Prob. solving practices	−0.264	0.031	0.164	−0.071	−0.071	0.084	0.000	0.000	0.000	0.000	0.000
Student-centered practices	−0.067	−0.351	0.055	0.071	−0.016	0.035	0.000	0.000	0.000	0.000	0.000
Classroom instruction	0.046	−0.071	0.049	−0.040	−0.053	−0.072	0.000	0.000	0.000	0.000	0.000
Achievement											
Achievement 1995–1999	0.035	−0.095	−0.186	0.113	−0.142	−0.084	−0.473	−0.013	−0.087	0.099	−0.038
Achievement 1995–2000	−0.025	−0.148	−0.206	−0.016	−0.124	−0.103	−0.642	−0.006	0.002	0.075	0.038
Achievement gap											
Gender gap 1995–1999	−0.091	−0.203	−0.209	−0.054	−0.172	0.018	−0.549	−0.006	0.002	0.056	0.014
Gender gap 1995–2000	−0.062	−0.159	−0.360	0.025	−0.152	−0.017	−0.586	−0.071	0.042	0.063	−0.068

result in more modest gains in achievement levels for the 1995–99 period (−.19 SD) and 1995–2000 period (−.21 SD). Finally, the strongest relationships were obtained for instructional practices, in which a one-standard deviation increase was associated with a .47 or .64 standard deviation decrease in achievement gains across the two time periods.

In contrast to the results from the Miami-Dade site, the analysis of the achievement gap indicators (6.1 and 6.2) suggest relationships that are very similar to those of general achievement gains. Specifically, the three indicators (influence on policy, teacher preparation, and instructional practices) emerge as the most influential in explaining variability among the gap changes, and all of the substantial relationships were in an inverse direction. The path coefficients ranged from −.16 (for influence on policy) to − .59 (for instructional practices).

Many of the findings presented in our analyses of Miami and Chicago show disappointing results from the perspective of the reform. Not all teachers and students experienced positive impacts from the reforms. Recall in chapter 5 that approximately 25 percent of all of the teachers in our study consistently used student-centered classroom instructional practices that ultimately fostered student achievement gains. The reader should also keep in mind that when teachers referred to "preparation" in our analyses presented here, they had in mind gaps in their preparation. This may help to explain the findings reported here that show positive influences of standards-based professional development on classroom instruction, but negative or modest gains in student achievement outcomes. There is evidence that USI reforms are in fact having a positive impact and simply require more time to be manifest in student achievement gains. In addition, the measures of student achievement used here are district's own standardized, mandated high stakes tests, not the most valid measures, perhaps, of standards-based teaching and learning.

Seven-Driver Model

The seven-driver model (which included a driver related to school culture) explained slightly more of the variability in overall achievement (22.4 percent) and variability in the achievement gap (27.8 percent) than were observed with the six-driver model. The total effects estimated from this model are presented in Table 7.8. The inclusion of the seventh driver did not substantially reduce the path coefficients observed in the six-driver model (Table 7.7), with influence on policy, teacher preparation, and instructional practices evidencing inverse relationships with both

Table 7.8.
Seven-Driver Model: Total Effects of Exogenous Variables (Policy, Resources, Stakeholders, Culture, and Instruction) on Endogenous Variables (Achievement and Instruction) for Chicago

Endogenous Variables	Exogenous Variables												
	Policy		Resources		Stakeholders		Culture		Instruction				
	Driver 2.1 Views of Reform Impact	Driver 2.2 Influence on Policy	Driver 3.1 Teacher Preparation	Driver 3.2 Professional Development	Driver 4.1 Mission	Driver 4.2 Assessment Plan	Driver 7.1 School Culture	Driver 7.2 Teacher Opinions	Driver 1.1 Instructional Practice	Driver 1.2 Prob. Solving/ Technology	Driver 1.3 Prob. Solving Practices	Driver 1.4 Student-Centered Practices	Driver 1.5 Classroom Instruction
Instruction													
Instructional practice	0.052	0.169	0.602	0.316	0.019	−0.003	−0.053	0.083	0.000	0.000	0.000	0.000	0.000
Prob. solving/tech.	−0.039	−0.215	0.117	0.182	0.173	−0.071	−0.032	0.287	0.000	0.000	0.000	0.000	0.000
Prob. solving practices	−0.237	0.048	0.065	−0.131	−0.027	0.102	−0.061	0.161	0.000	0.000	0.000	0.000	0.000
Student-centered practices	−0.021	−0.337	−0.096	−0.037	0.033	0.050	−0.032	0.287	0.000	0.000	0.000	0.000	0.000
Classroom instruction	0.090	−0.059	−0.102	−0.147	−0.004	−0.057	−0.032	0.287	0.000	0.000	0.000	0.000	0.000*
Achievement													
Achievement 1995-1999	0.044	−0.136	−0.158	0.090	−0.201	−0.122	0.182	0.063	−0.474	−0.094	−0.085	0.119	−0.021
Achievement 1995-2000	−0.030	−0.189	−0.132	−0.003	−0.194	−0.143	0.181	−0.031	−0.610	−0.037	0.018	0.105	0.059
Achievement gap													
Gender gap 1995-1999	−0.086	−0.225	−0.191	−0.065	−0.205	−0.003	0.101	0.030	−0.547	−0.048	0.003	0.065	0.022
Gender gap 1995-2000	−0.082	−0.191	−0.252	0.074	−0.219	−0.050	0.135	−0.129	−0.546	−0.066	0.061	0.089	−0.051

overall achievement and changes in the achievement gap. However, indicator 7.1 of the culture driver (school culture) evidenced a direct relationship with achievement gains and with changes in the achievement gap. Specifically, a one-standard deviation increase in school culture is expected to result in a .18 standard deviation increase in achievement change. Similarly, a one-standard deviation increase in school culture is expected to result in a .10 to .14 increase in the gap indicators (6.1 and 6.2). These are important influences and can be addressed by policy at the school level to strengthen the school culture by promoting teacher engagement in learning communities, as we describe in chapter 9.

SUMMARY

Substantial differences were observed in the results obtained from the two sites included in these analyses. Because the principal components extracted from the samples in the Miami-Dade and Chicago sites suggest conceptually different indicators being generated for the path models, a simple one-to-one comparison of path coefficients is neither possible nor desirable. From a broader view, however, the data from both sites suggest that differences in instructional practice are statistically related to differences in achievement (both general achievement and achievement gaps). Further, both sets of data suggest that policy and resource applications are measurably related to student outcomes (albeit by different mechanisms). Finally, both sets of data suggest that differences in school culture contribute to the explanation of achievement differences above and beyond the explanatory power provided by the other drivers.

Interpretive Caveats

The results of our quantitative analyses must be interpreted within the limitations of this study. The most important limitation of the analyses presented in this study is that of the sample size obtained from each of the sites. The small samples necessitated the use of statistically simpler path analysis, rather than the more complex (albeit more rigorous) approach of structural equation modeling. The validity of path analytic results is dependent upon the tenability of assumptions and these assumptions are difficult to evaluate in small samples. For example, departures from linearity or additivity may be rigorously tested in large samples, but such tests are woefully lacking in statistical power when the sample size is small.

Random measurement error is unlikely to present a problem in these data because of the use of component scores rather than observed variables (the first principal components tend to be relatively free of random measurement errors). However, the potential for model specification errors in the form of omitted variables should not be disregarded. Although multiple data sources were brought to bear in these analyses, additional variables that impact the drivers we have investigated hold the potential to bias estimates of path coefficients. Finally, the assumption of recursive relationships (a one-way causal flow) is important. The complexity of school and classroom processes suggests that reciprocal relationships may be anticipated. The investigation of such relationships requires a longitudinal research design, and the collection of school and classroom process measures at multiple points in time.

Future Research

In considering recommendations for methodological approaches appropriate for conducting future research that examines the impact of reform on student outcomes, several caveats are in order. First, multiyear, multisystem studies are expensive to undertake and require the development of trusting relationships between school districts and research teams. To address this concern, consortia of researchers interested in education reform might agree on important variables, and collaboratively, develop and test models using data from the school systems in which they have forged relationships using a meta-analytic approach. Absent such an undertaking, individual researchers might conduct multilevel analyses that acknowledge relationships among levels such as students nested within classrooms (individual versus group achievement), classrooms within schools (school achievement), and schools within urban districts (saturation of reform initiatives). The decomposition of variability into its within (class, school, or district) and between (classes, schools, or districts) components is needed to estimate the impact of reform, which may have institutional or collective effects as well as individual effects on student outcomes (Bidwell and Kasarda, 1980; Gamoran, 1984, 1987). However, the current project was hampered by the costs associated with more a desirable repeated-measure, multiyear design.

In sum, future research could address the following concerns and difficulties: (1) limited inferences from small samples; (2) reliance on teacher self-report survey data; (3) linking classroom observations with achievement data; (4) exploring the link between preparation, professional development, and the quality of the school's culture in increasing and maintaining student achievement outcomes; (5) examining teacher and student reports of teachers' standards-based practices in mathematics and science classrooms armed with contextual knowledge about the policy environment in which they work; and (6) developing longitudinal databases which focus on changes in schools, teachers, and students over time. Future studies might include measures of the perceived policy environment, teacher preparation, and professional development as well as the teaching practices within the school (Cohen and Hill, 2000; Desimone et al., 2002). The next steps for examining the impact of reform initiatives on student achievement fall into three broad areas of concern, as outlined below and discussed next.

Hierarchical Data and Small Samples

Efforts to increase confidence in future results might include the use of a nationally representative sample of students, teachers, principals, schools, and districts, which would increase sample size and improve the generalizability of findings to the national landscape. Even the most ambitious sampling plans fall victim to attrition when our goal is multiple measures at multiple time points. Students and teachers move around districts, and schools are living systems always undergoing change. Cross-sectional studies may yield more desirable sample sizes but limit our ability to evaluate causality. Complete data on a larger number of classrooms in each district or the same measures across districts would have facilitated improved analysis strategies in the current study. As mentioned earlier, small samples bias estimates, causing one to question the meaning of the results. School-level analyses may avoid this concern but answer different questions about the impact of the reform. Increasing the sample of classrooms within schools—in other words, observing and surveying a minimum of twenty teachers and their students in a school—would increase our confidence, but increasing the sample size at the district level limits us to larger systems, those with a hundred or more schools. Simulation studies suggest balanced groups, the number of groups being more important than the size of the groups.

Measurement Issues

Of particular interest is the ability to examine the preparation and reported practice of urban teachers. Future research might connect student achievement, teacher preparation, teacher practice, and school culture variables so we may examine directly the impact of policy on practice and its indirect effect on student outcomes. For example, using the School and Staffing Survey (SASS) to identify high-reform schools through principal reports of performance standards for students (such as creating composites to measure policy influences and progress) allows the comparison of schools in terms of their self-reported progress and its relationship with student outcomes. The Principal Questionnaire from the SASS affords the opportunity to link what principals indicate happens in their school with what their teachers report doing in that school. Previous work (Kersaint, Borman, Lee, and Boydston, 2001) attempted to make this connection through principal interviews and teacher surveys. Since these items were answered in a national sample of schools, we should be able to identify within districts a larger sample of high-reform schools. Locating schools on a continuum from low to high also permits the study of relationships among schools reporting progress toward their performance goals.

Professional development items from the SASS Teacher Survey and the SASS Principal Questionnaire could be used to determine the professional development climate in schools. These items document teacher perceptions of the importance of professional development activities, the purpose of these activities, and the teacher role in selecting activities. In addition, these instruments can be used to develop a profile of participating schools by linking the demographic information available in the School District Questionnaire and the Public School Questionnaire. Each of these instruments adds information about the policy context in which school reform takes place. Unfortunately, the SASS does not include student achievement data and therefore linking to other databases will be necessary.

The National Assessment of Educational Progress (NAEP) is a nationally representative assessment of students' knowledge and achievement in several subject areas, including mathematics and science. NAEP questionnaires ask participating students to respond to questions regarding course instruction. The NAEP assessment gathers information regarding student instructional experiences and school environments that are of interest in exploring the correlates of achievement for minority students in large urban school districts. Teachers are asked about their professional preparation and instructional activities, and school principals respond to queries about practices and policies affecting their school. Additionally, because the mathematics assessment in 1996 was closely tied to the NCTM standards' five content strands, the NAEP data are useful in investigating student achievement in those districts that were engaged in the NSF reforms during the years 1993–1996. The NAEP uses a multistage design to select a nationally representative sample of students that reflects students in the nation's schools and requires special techniques to obtain unbiased estimates because of the complex sample design. By connecting across these surveys we might satisfy our need for valid measures without creating another national database.

Use of Longitudinal Data

Our attempt to connect teacher practices and student achievement was hindered by privacy concerns that limited our access to student-level data. Future studies need access to individual student data that can be linked to their teachers. The multilevel character of school research must recognize that, while all teachers within a school are housed in the same policy environment, individual teachers' preparation, in-service activities, and practice may result in variable outcomes for students. At the same time, students are mobile both within schools (promotion and retention) and across schools (mobility). Information about individual student growth is necessary to understand the impact of policies on both teacher practices and student achievement outcomes. In order to understand educational change, both antecedents and outcomes must be measured at multiple time points using valid and reliable measures. We are interested in effects at multiple levels: district-level analyses viewed as systemic effects may be assessed, school effects within individual districts viewed as saturation of the reform effort, and, finally, teacher effects to determine how individual teachers embedded within a policy web of systemic reform are able to effect differential student outcomes for similar groups of students. These relationships may be modeled both for all students and for subgroups of students since teacher practices that may not appear to increase achievement for all students may in fact improve achievement outcomes for particular subgroups (such as African Americans or Latinos), but this requires repeated measures of performance of the actors at each level.

NSF funded a number of urban school districts in its effort to improve achievement outcomes for all students attending schools in the nation's most economically strapped school systems. There will continue to be interest in the academic achievement outcomes of all students, but especially those students in urban districts. Policy makers need information about the long- and short-term effects of policy mandates. Reform is synonymous with education; perhaps the time has come to model reform effects using up-to-date yet nationally representative longitudinal databases.

CONCLUSION

The current thrust for systemic reform addresses the assumption that schools have not provided students, especially those students attending the least academically successful schools, with knowledge necessary to be successful in society—the outcome stressed by Newmann and Wehlage (1995). Students were not sufficiently challenged by the instruction they received, with the result that many were ill prepared to attend college upon graduation, enter technologically complex careers, or engage in challenging intellectual work. By setting more rigorous standards for students, the general level of student achievement would rise, better preparing students for post-secondary educational opportunities and employment (Roeber, 1999). Our overall analysis of mathematics achievement of the schools in our study indicates the achievement gap has been reduced. In addition, our analyses using comparison non-USI sites show that, compared to relatively affluent, predominantly white schools, USI schools have made great progress in closing the gap. These analyses also suggest that it is difficult to sustain an upward trend in achievement at the middle school level.

Our results add texture and complexity to our knowledge of classroom reforms, showing, for example, that factors such as technology in the classroom have a strong impact on achievement for many Latino and white students but may not leverage similar gains for blacks and males. Indeed, more research is needed to examine this phenomenon. We also learned from these analyses that students whose teachers use performance projects, technology projects, and standards-based practice experience gains in mathematics achievement. In addition, community stakeholders' involvement in schools positively affects student achievement. An obvious policy lesson here is that poorly performing schools must garner support from various constituencies to improve student outcomes. Our models indicate that another important aspect of improved student outcomes is the school culture or learning community environment. In schools where teachers view themselves as learners and believe that their students *can* achieve, improved student outcomes are likely to result. The data examined from the four sites consistently underscore the importance of classroom processes and the critical role of teachers in successful reform efforts.

8

School Culture: The Missing Lever in Improving Student Outcomes and Achieving Sustainable Reform

Improving student outcomes is a complex enterprise linking teachers' professional development activities and their subsequent classroom practices. Previous research is clear that professional development is indeed an effective lever for improving classroom instruction. Many studies have shown that when teachers are provided long-term standards-based professional development, their practices are much more likely to engage students in problem-solving activities, in active learning, and in other forms of instructional activity likely to lead to enhanced student learning and interest (National Research Council, 1996).

The results of our analyses of student data in chapter 7 were somewhat disappointing because we did not see consistent linkages between teachers' professional development, classroom instruction and student academic achievement gains in mathematics. Although the reforms influenced many teachers (at least one-fourth of those in our research) to engage in student-centered, standards-based practices, not all teachers did so contributing to the negative findings discussed in chapter 7. The reforms, in other words, were not systemic in that all teachers did not adopt student-centered practices. But, what if we considered the role of school context or culture as a moderating factor?

In this chapter we argue that a critical mediating variable in this process is the school context and, specifically, a professional community of practice that sustains teaching and learning in line with the reforms. The push for higher standards and increased student achievement, as well as policies such as the National Science Foundation's Urban Systemic Initiative aimed at reforming entire urban school systems in the United States, has emphasized the importance of professional development for teachers (Borko, Mayfield, Marion, Flexer, and Cumbo, 1997). Increasingly, researchers have also argued that communities of practice at the school level are critical to sustain reforms and encourage the development of instructional approaches supporting student achievement (Lee, Smith, and Croninger, 1995). We maintain that for reform to take hold in schools and classrooms it is not enough simply to mandate more hours of professional development, no matter how well designed. We believe that support at the school level is critically important to sustain reform and support lessons learned in professional development.

Our conversations with school principals, as shown in chapter 2, revealed that many of them viewed facilitating the professional development of instructional staff as a key responsibility. In other words, principals endorsed the importance of school culture as critical to understanding links among professional development, classroom practice, and gains in student achievement. The analyses we present in this chapter are related to a number of issues discussed in earlier chapters. For example, in chapter 4 we examined the impact of professional development as reported by teachers in our project, while chapter 5 examined the range of classroom practices (subject-centered, teacher-centered, and learner-centered) that we observed and their relationship to student achievement. In chapter 7 our analyses of multiple indicators informed our study of NSF's six-driver model of systemic reform using path analysis to test a model that included an additional seventh driver: school culture.

As discussed in chapter 7, school culture explained 20 percent of the variance associated with gains in student achievement in our Miami-Dade school sample. Our related analyses of school culture components from the Survey of Classroom Practices (SCP) had also demonstrated that a change in teachers' opinions regarding a culture of sharing and their beliefs about standards resulted in gains in mathematics achievement. Additionally, a change in the measured school culture decreased the achievement gap in two USI sites. In this chapter we use school-level achievement data to answer questions about the relationships among professional development, classroom practices, and school culture and the link to achievement. First we turn to a discussion of how we conceptualize school culture in our work.

SCHOOL CULTURE

Organizational culture, according to Peterson and Deal, is "the underground stream of norms, values, beliefs, traditions, and rituals that has built up over time as people work together, solve problems, and confront challenges" (1998, p. 28). Schools are social organizations with their own structures, norms, belief systems, and hierarchies of roles and responsibilities. We have seen in chapters 5 and 6 that how a school organizes its classroom schedules can make a difference in student outcomes. Classrooms in schools that used block scheduling, as did many of our schools in El Paso and Miami-Dade, were less effective in organizing instructional time in promoting student engagement.

A strong, nurturing culture within a school fosters the development of teacher leadership, and in turn should produce positive results in student outcomes. Given the importance of relationships, Hargreaves and Fullan (1998) suggest that highest priority be placed not on restructuring but rather on "reculturing" the schools. The goal of reculturing the school is to engage teachers and other stakeholders to work together differently by creating more collaborative work cultures. In their guidelines for principals they suggest attention to emotional management "which is ultimately about attending to the relationships within the school properly" (p. 119). Evidence from several studies, including those undertaken by Newmann and Wehlage (1995) and Louis, Marks, and Kruse (1996), supports the notion that students learn more productively when principals, teachers, and others work collaboratively within a professional learning community.

If standards-based practices are to take hold and become the mainstay of teachers' instructional repertoires, more is necessary than simply logging time in professional development. What matters most is that teachers be given the opportunity to spend time with colleagues in a supportive setting, learning from their peers and practicing with them strategies that allow students to take an active role in their own learning. To determine the extent to which teachers worked in a school culture that valued and nurtured systemic changes in mathematics and science instruction, we arranged for all members of the school workforce (not including the principal and his or her staff) to complete the School Culture Quality Survey (SCQS). The SCQS defines professional community using ideas and concepts from the quality movement and the work of Peter Senge and colleagues (1990, 1994, 2000). The SCQS defines a desirable professional community as including a shared vision of the school's future, a cadre of colleagues who view their work as actively facilitated by their leaders, and who work together productively toward common goals, actively seeking and learning together the new skills and knowledge needed to achieve the desired school future.

Perhaps even more critical to the process of standards-based reform is the orientation of the school principal to the creation and maintenance of a nurturing school culture. Student-centered leadership at the school level in part rests upon the ability of the principal to nurture the capacity of school organization members to surface, critique, and apply mental models that people bring to the world (Senge, 1990, 1999). We next identify the components of school culture (shared vision, facilitative leadership, teamwork, and learning communities) that informed our research.

Shared Vision

Peter Senge (1990), after studying many of the most successful business organizations in the United States, identified five "disciplines" or characteristic ways of behaving shared by highly successful organizations. The disciplines of "shared vision" and "team learning" were viewed as continuing ongoing processes, shared by leaders and team members. The leader's responsibility is viewed as facilitating the continuing development of the vision with team members, and serving as custodian of the vision, articulating and rearticulating its evolving message and refining and adapting it with team members. In his *Fifth Discipline Field Book*, Senge (1994) argues, "choosing to continually listen for that sense of emerging purpose is a critical choice that shifts an individual or a community from a reactive to a creative orientation." Kouzes and Posner (1987) state that effective leaders hold in their minds visions and ideals of what can be. But visions seen only by the leaders are insufficient to create organized movement. Others must also see exciting future possibilities.

Facilitative Leadership

Thomas (1992) asserts that leaders at the school site have a significant bottom-up role to play in making changes in teaching and learning. Principals set the context within the school. The importance of school culture in this dynamic cannot be overlooked. If a positive culture exists in a school and fosters teacher leadership, positive results with respect to student outcomes are produced. Resnick and Hall (1998), in their work in reforming instruction in schools, adopted a nested learning community approach and found that teachers find their primary community of learners among their peers; however, it is the interaction between role groups that constitutes the nesting feature of learning communities. Thus the orchestrator of the school-based learning community for teachers is the school principal.

Teamwork

Teamwork, as separate from the learning community or team learning, has not been examined extensively in recent literature. In our factor analytic studies of the items gathered in the development of the SCQS a cluster of items that were clearly identified with that concept continued to appear as a separate cluster from those with content integral to the notion of learning community. Teamwork, then, signals the capacity of the organization's members to view themselves constituting a group wedded to collaboration and intensive cooperation across grade levels and classrooms. The differentiation of this concept from the learning community reflects that it is possible to work together in the interests of the children without learning together. Teaching can be the loneliest of professions, and many teachers in many schools unfortunately work behind closed doors (Lortie, 1975).

Learning Communities

Louis, Marks, and Kruse (1996) note that, while enhancing individual teacher skills and knowledge is important, attention must be paid to the development of professional learning communities, described as teachers' collective engagement in sustained efforts to improve practice. Resnick and Hall (1998) advocate nested learning communities where not only students but all education professionals are learners. Teachers, principals, and central office administrators form communities of adult learners who are focused on improving their practice. Schools become places where learning is the work of both students and professional educators and where continuous learning in pursuit of educational improvement is the norm. Resnick and Hall define professionals as individuals who are continually learning rather than as people who already know.

Rosenholtz's (1989) research identified "learning enriched" schools—places where students were achieving—and "learning impoverished" schools—places were students were not achieving. She found the key difference to be whether teachers were involved in collective learning and sharing about the practice of teaching. A collaborative environment such as this has been described in the literature as a learning community. Lieberman (1996) described a learning community as a community where teachers and administrators share and discuss their work experiences, contribute to and gain access to learning that solves immediate problems of practice, and grapple with problems in greater depth and complexity.

Boyd and Hord (1994) found that building a learning community in the school reduces isolation, increases staff capacity, and provides a caring, productive environment. Lieberman (1992) cites powerful effects on students, on the culture of the school, and on teachers' sense of efficacy when they are part of a learning community. Barth (1988) suggested that every teacher be encouraged to get involved in school change efforts through being part of a learning community because, as they learn and grow, they influence teachers around them. Indeed, it is only being invested in the culture of the school that classroom practices and teaching and learning can be engaged. Such engagement requires organizational supports at the school level, as we discuss next.

INFLUENCE OF SCHOOL CULTURE ON CLASSROOM PRACTICES

Teacher support typically takes the form of mentor teachers and others who join with classroom teachers to provide communities of collaborative, reflective practice (Stein, Silver, and Smith, 1999). Communities of practice—teams of teachers working together over time at the school level—appear to be critical in implementing classroom and school-specific standards-based instructional approaches in mathematics and science. Teacher teams of this sort are likely to be most effective when teachers are provided adequate standards-based curricular resources, sustained classroom-based coaching, and the time to pull it all together (Putnam and Borko, 2000).

NSF's efforts to enhance student outcomes in mathematics and science rest upon intensive and sustained professional development that, ideally, provides teachers with direct, hands-on, problem-solving pedagogical strategies and content knowledge to inform and sustain standards-based classroom practice. In Chicago and El Paso this was accomplished by sustained, long-term teacher support, and opportunities for degree-seeking course work at local universities.

From previous research we know that some specific programs designed by university-based researchers and their colleagues have positive outcomes for teachers' classroom practice. In mathematics the Cognitively Guided Instruction program (Carpenter, Fennema, Peterson, Chiang, and Loef, 1989) and in science Project Based Science (Krajcik, Marx, Blumenfeld, Soloway, and Fishman, 2000) are designed to promote inquiry and student learning using technology tools. These programs and others like them emphasize approaches advocated by Putnam and Borko, (2000) as well as other researchers and practitioners who view cognition as situated. This perspective underscores the importance of authentic activity complemented by intensive learning in a classroom context of social and shared experience. Involvement in hands-on, problem-solving activities by students in mathematics and science at all grade levels should lead to greater student achievement, according to these researchers.

Communities of practice incorporating teams of teachers are critical both to developing classroom and school-specific standards-based instructional approaches in mathematics and science and to increasing student achievement. USI projects employed programs of intensive professional development focused on constructivist strategies for teaching mathematics and science across grade levels K–12. While teacher professional development is the primary vehicle used for improving classroom instruction, our research indicates that individual schools create a context within which professional development facilitated teacher and student learning and that this makes or breaks professional development's ultimate impacts on both instruction and student advancement outcomes.

A number of professional development strategies from the literature support our view that a school-based community of practice enhances classroom instruction and student achievement. These include: (1) providing professional development in the school (endogenous professional development in our lexicon) (Borko, Mayfield, Marion, Flexer, and Cumbo, 1997); (2) using teaching cases generated by teachers (Barnett, 1998; Barnett and Friedman, 1997); (3) creating and maintaining an atmosphere of support and enhancement (Campbell and White, 1997); and (4) using mathematics and science specialists and master teachers as resources (Le Tendre and Chabran, 1998).

We use the term *exogenous* to characterize professional development activity that emphasizes formal training from outside experts and requires external incentives for teacher participation. We use the term *endogenous* to characterize professional development activity that views teachers' day-to-day practice and student work as major components of the curriculum, abetted by teachers' intrinsic motivation to work as experts and learners in their school-based learning communities.

The work of Spillane (2002) suggests that district officials operating from a behaviorist perspective are not effective in supporting teachers' implementation of standards-based reform, in contrast to those guided by a situated or cognitive perspective who saw teachers as agents of their own learning in much the same way that standards-lead mathematics and science reform views students' roles in learning requiring *their* active engagement. Unfortunately, behaviorist-oriented administrators comprised 85 percent of those Spillane interviewed in his research. In contrast, district change agents who supported a situated or cognitive perspective saw teachers as active agents in their learning enabled through reflection. These district officials viewed teachers' day-to-day practice and student work as major components of the curriculum abetted by teachers' intrinsic motivation to work as experts and learners in their educational communities. Differences between the two distinct administrator conceptualizations of professional development identified by Spillane are apparent with respect to the locus of responsibility for professional development.

MEASURING SCHOOL CULTURE, PROFESSIONAL DEVELOPMENT, AND CLASSROOM PRACTICES

The SCQS served as the measure of school culture and was completed by teachers and support staff in participating schools. The SCQS scales are shared vision, facilitative leadership, teamwork, and learning community. The content of the scales reflects: (1) the extent to which faculty and support staff see themselves as participants in a *shared vision*; (2) the extent to which they view the leadership of the school to be actively facilitating their work (*facilitative leadership*); (3) the extent to which they view themselves as working together cooperatively and effectively in an atmosphere of mutual respect and caring (*teamwork*); and (4) the extent to which they perceive that teachers, support staff, and members of the school's administration are eagerly investing, learning, and working together toward realization of the vision and goals they share (*learning community*). The alpha reliability for each of the scales of the SCQS varied from .90 to .93. The SCQS forms were distributed to teachers and other staff at each of the forty-five participating schools. The purpose of the surveys in this series of analyses was to inform our understanding of the links among mathematics achievement, school culture, instructional practices, and professional development activities. The first step was to determine the variance in students' mathematics achievement gains explained by information about the classroom practices and professional development activities of the mathematics teachers, and the school culture at each of the participating schools. Professional development (PD) has been a focus of the NSF-sponsored efforts to improve student achievement in mathematics and science.

Items from the teacher form of the Survey of Classroom Practices (SCP) provided a measure of teachers' participation in professional development activities. Teachers responded to twelve professional development items that asked them to reflect on the amount and impact of these activities on their classroom practice. Teachers' reports of their classroom practices formed a second predictor of student achievement. Fifteen items from the SCP that asked teachers to report on how often students in their class were engaged in a particular instructional activity were selected for further analysis. We then linked the school-level variables from the SCP, school-level mean mathematics achievement scores, teachers' reports of their professional development experiences, and results of the SCQS.

Potential links between students' mathematics achievement gains and the school culture as measured by the SCQS formed the third area for analysis. Our criterion of interest was student mathematics achievement, using school grade-level achievement reports from the *National School-Level State Assessment Score Database* (NSLSASD) and summaries of student-level achievement data provided by cooperating districts, which were then used to develop a school-level students' mathematics achievement gains score. Our final criterion of mathematics achievement gains was realized by using school-level achievement data to determine achievement levels at baseline and at the end of the project period. We controlled for poverty using the school's percentage of students on free/reduced lunch and the mobility rate. Before turning to the results of our analyses of school culture and its impact on achievement, we will present an important initial discussion of student outcomes using the NSLSAD to understand how USI schools differed from schools in districts not participating in the NSF reforms.

Did USI Make a Difference?

In order to answer the question of whether or not USI reforms had an impact on student outcomes, we looked at NSLSASD data to see how well USI schools performed in comparison to schools in districts nearby but not involved in USI reform efforts. These districts were in fact generally more affluent than districts receiving support from NSF. Our earliest analyses of these data pointed to a closing of the achievement gap. Figure 8.1 charts the progress or the gains produced by USI schools during the period of NSF support.

In making our comparisons, we first combined data across school level (elementary, middle, and high school) within district and then completed what could be termed an "omnibus" comparison between

Figure 8.1. Standardized Math Mean Scores for USI and
Non-USI Schools

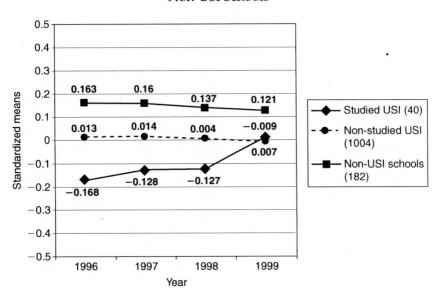

our USI and non-USI schools. Our analysis compared forty studied USI schools and 182 non-USI schools. Note that the dotted line represents USI schools not included in our study. Because USI grants were made to participating districts, one could consider all schools in our four districts as USI schools. Thus, the dotted line represents all schools in each district, less our studied USI schools. In Figure 8.1 we see that our studied USI schools performed at a level that was more than one-third of a standard deviation below non-USI schools at the baseline. However, over four years of the USI reforms, the USI schools closed that gap to just over one-tenth of a standard deviation. Our repeated measures analysis of variance yielded no main effect of time, and a borderline main effect for group (studied USI versus non-USI) $F = 3.56$ (1,220), p. = .06. Time by group interaction was also a borderline effect, $F = 2.15$ (3,660), p = .10. Further analyses revealed that differences between groups in 1996 and 1997 were significant ($p < .05$), while the differences between the groups in 1998 and 1999 were not, indicating movement toward closing the gap. It is also reasonable to assume that there would be differences in the gap and changes across time by school level. We offer the data in Figure 8.1 to assure our reader of improvements in student achievement before we present evidence of the importance of school culture.

MEASURING STUDENT ACHIEVEMENT, PROFESSIONAL DEVELOPMENT, AND CLASSROOM PRACTICES ACROSS SITES

We next turn to a discussion of the achievement measure used in the next set of analyses. Scores on district-administered mathematics achievement tests during the first and final years of the USI study were used as the measure of mathematics achievement. While schools within each participating site used the same instrument for measuring mathematics achievement at various grade levels, measures of mathematics achievement varied among the districts. To make mathematics achievement scores comparable across districts, within-district school-level mathematics achievement gains were converted to z-scores (mean of zero and standard deviation of one) based on the within-district mean and standard deviation. Comparability of the resulting scores rests on the assumption that, although different instruments were used, each instrument is an adequate measure of the same construct (mathematics achievement). The individual schools varied significantly with respect to such variables as percentage of students on free/reduced lunch, pretest scores, and mobility rates. These variables were already established before the study began but are known to explain significant amounts of the variability in school mean mathematics scores. For this reason, the final-year school-level mathematics achievement z-scores were predicted from (1) first-year mathematics achievement scores, (2) percentage of students on free/reduced lunch, and (3) mobility rate of the school. Differences between the actual and predicted school-level mathematics achievement scores were used as the criterion for gain in mathematics achievement (students' mathematics achievement gains).

We created mean scores for individual SCP items, matching items for each version of the classroom practices survey (mathematics, science) to create school scores that included both mathematics and science teachers. Table 8.1 presents the means and standard deviations for eleven SCP survey items related to teacher professional development activities. In addition, the relationship between each item and our criterion measure is presented. Using linear regression and factor analysis, a subset of items was identified that formed two components of teacher-reported professional development activities. These two components describe professional development activities delivered to teachers outside their schools (exogenous) and professional development activities chosen by teachers within a school to build their professional community (endogenous). The standardized regression coefficients from the Promax rotated factor pattern are presented in Table 8.2.

Table 8.1.
Professional Development Items, Means,
Standard Deviations, and Correlation with Students' Mathematics
Achievement Gains

Professional Development Emphasis	M	SD	Students' Mathematics Achievement Gains	Probability
Implementing content standards	1.6880	0.5734	−0.126	0.403
Implementing new curriculum or instructional materials	1.8275	0.5594	−0.220	0.142
New methods of teaching	1.7529	0.5661	−0.145	0.335
In-depth study of math content	1.3620	0.5494	−0.105	0.487
Meeting the needs of all students	1.7015	0.5376	−0.086	0.571
Multiple assessment strategies	1.7504	0.6449	−0.059	0.696
Network or study group on improving teaching	1.0083	0.6804	0.219	0.144
Formal portfolio assessment	0.6627	0.6736	−0.024	0.873
Extended institute or professional development	0.7569	0.7219	0.139	0.356
Observed other math teachers	0.5471	0.5822	0.115	0.448
Read or contributed to professional journals	0.9692	0.6920	−0.146	0.333

Table 8.2.
Professional Development Factors

Professional Development Emphasis	Exogenous Professional Development	Endogenous Professional Development
New methods of teaching	88*	−9
Implementing content standards	86*	−8
Multiple assessment strategies	85*	−5
Implementing new curriculum or instructional materials	74*	5
Meeting the needs of all students	68*	11
In-depth study of math content	63*	25
Formal portfolio assessment	20	67*
Network or study group on improving teaching	−10	63*
Observed other math teachers	−6	62*
Read or contributed to professional journals	22	35
Extended institute or professional development	27	30

* = $p < .01$.

Table 8.3.
Classroom Practices, Items, Means, and Standard Deviations and
Correlation with Students' Mathematics Achievement Gains

Classroom Practices	M	SD	Students' Mathematics Achievement Gains	Probability
Use performance tasks	2.0920	0.6712	0.285	0.055
Use individual or group presentation	1.6391	0.6599	0.330	0.025
Teach LD strategies	1.6518	0.4376	0.307	0.038
Select or adapt inst materials	1.8768	0.4611	−0.214	0.154
Assignment longer than a week	1.2641	0.6461	0.203	0.177
Work with group to improve written work	1.3112	0.6060	0.248	0.097
Review for test or quiz	1.4210	0.5595	0.175	0.245
Show required steps	1.9562	0.5183	−0.289	0.052
Explain reasoning or thinking	1.6819	0.5029	−0.353	0.016
Work on demo project or proof	1.4761	0.5623	−0.209	0.164
Use sensors or probes	0.4536	0.5830	0.250	0.094
Collect data using the Internet	0.9185	0.6137	0.199	0.185
Use portfolios	1.0779	0.7642	0.121	0.423
Use manipulatives	1.8196	0.9563	−0.006	0.971
Take a quiz or test	1.4446	0.4237	−0.128	0.396

Next, focusing on teacher-reported instructional activities, we applied the same strategy of combining linear regression and factor analysis to develop three measures of instructional practices. The means, standard deviations, and correlations of these items with the mathematics achievement gain criterion are presented in Table 8.3, and the standardized regression coefficients from the Promax rotated factor pattern are presented in Table 8.4.

We determined that the fifteen instructional practice items identified as having a strong, positive relationship with students' mathematics achievement gains suggested a three-factor model. The first factor was assigned the name *standards-based instructional practices* and subsumes items that measure what can be seen as authentic teaching practices such as use of portfolios, sensor probes, group work, collection of data via the Internet, individual or group presentations, and assignments taking longer than a week to complete, with reviews for tests and quizzes.

The second factor, assigned the name *standards-based performance practices*, subsumes items that involve performance tasks, use of manipulatives, the use of strategies for the learning disabled, and selecting or adapting instructional materials with an eye toward alternative forms for understanding and evaluating students' abilities. The third factor, assigned the name *homework practices*, subsumes items related to homework tasks such as demonstrations or proofs, explanation of reasoning or thinking, showing required steps, and taking a test covering homework. These tasks (for example, homework assignments) are undertaken by students for the most part without guidance from their teachers. In other words, these are practices that are neither carried out in the presence of teachers nor monitored by teachers. It is important to note that only the *homework practices* factor related negatively to students' mathematics achievement gains. Teachers' involvement in instruction is critical to students' mathematics achievement gains, according to our findings.

Linear regression was used to determine factors most clearly linked to students' mathematics achievement gains. For this purpose a series of regression analyses using the *standards-based instructional practices*, *standards-based performance practices*, and *homework practices* factor

Table 8.4.
Instructional-Practice Factors

Classroom Practices	Standards-based Instructional Practices	Standards-based Performance Practices	Homework Practices
Assignment longer than a week	85*	−23	5
Use sensors or probes	71*	−6	−13
Use portfolios	68*	9	6
Work with group to improve written work	65*	8	3
Review for test or quiz	62*	10	26
Collect data using the Internet	53*	25	−1
Use individual or group presentation	48*	38	−13
Use performance tasks	22	73*	−16
Use manipulatives	−1	67*	3
Select or adapt inst materials	−29	51*	40*
Teach LD strategies	9	44*	−14
Work on demo project or proof	15	15	78*
Explain reasoning or thinking	20	−13	63*
Show required steps	−17	−11	59*
Take a quiz or test	−1	−19	40*

* = p < .01.

scores representing classroom practices, the exogenous and exogenous factor scores as predictors based on professional development activity, and the four subscale scores of the SCQS as measures of school culture were completed.

The number of schools included in the analyses varies from forty-six to thirty-six as a function of data elements missing from individual schools (only thirty-six schools completed the SCQS). In the discussion that follows we report the adjusted multiple correlation coefficient (Adj R^2) as the most appropriate statistic for comparing across samples of differing size because it adjusts for differences in variables-to-cases ratio in estimating the proportion of variability in students' mathematics achievement gains that would be explained by the predictor variables in the population of similar schools. We next explore whether endogenous or exogenous professional development leads to stronger gains in mathematics achievement.

Endogenous vs. Exogenous Professional Development

Social scientists engaged in education research frequently apply statistical "models" in the analyses they undertake to understand complex issues such as the one before us here. In other words, we wished to test different theories to explain gains in student achievement. Model 1 presented in Table 8.5 shows the results of a multiple regression analysis

Table 8.5.
Regression Estimates School-Level Mathematics Achievement

	Model 1	Model 2	Model 3	Model 4
	Professional Development	Instructional Practices (IP)	School Culture (SC)	IP + SC
Intercept	0.000	0.000	−3.089	−3.366
Exogenous	−0.183			
Endogenous	0.152			
Standards-based instructional practices		0.189*		0.220
Homework practices		−0.272**		−0.177
Learning community			2.595**	2.128**
Teamwork			2.146**	1.662
Shared vision			−3.809***	−2.796*
Model adjusted r^2	0.002	0.175	0.209	0.266

Note: * = p < .08; ** = p < .05; *** = p < .01.

with mathematics change as the dependent variable and the endogenous and exogenous professional development variables as predictors. Examination of the estimates reveals that less that one percent of the variability in mathematics change is explained by the two professional development variables. This leads us to conclude that teachers' reports of professional development activities as assessed by the related items of the SCP are not useful in predicting positive changes in students' mathematics performance over the four years of the reform. We continue to have an interest in the dichotomy of exogenous and endogenous staff development activities. This continued interest rests on the notion that exogenous staff development that leads to endogenous follow-up activity in a school seems likely to be more effective than exogenous activity that does not also have provisions for subsequent endogenous professional development at the school as an outcome. Given the importance of school culture in affecting student achievement, this seems to us to be a reasonable idea.

Instructional Practice Variables

In addition to both student mathematics achievement gains and types of professional development engaging teachers, we also wished to determine the kinds of classroom instructional practices that teachers used and if standards-based practices lead to increased student achievement as the research of Newmann and his colleagues (1996) leads us to expect. Table 8.6 reveals correlations between each of the three scales and students' mathematics achievement gains of .32 (p < 033) for the *standards-based instructional practices* scale, .20 (p < .18) for the *standards-based performance practices* scale, and −.39 (p < .007) for the *homework practices* scale. Model 2 presented in Table 8.5 shows that nearly 18 percent of the variability in students' mathematics achievement gains is explained by both the *standards-based instructional practices* scale and the *homework practices* scale (p < .006). Both the *homework practices* and *standards-based instructional practices* scales made significant contributions to students' mathematics achievement. Because the *standards-based performance practices* factor did not explain a significant proportion of the variability in student outcomes, the factor was not included in subsequent analyses.

Recall that we consider ineffective those practices related to unsupervised homework tasks that some teachers routinely assign to their students. As shown in our analyses, the *homework practices* scale correlates negatively with students' mathematics achievement outcomes. The *homework practices* items reflect the amount of time teachers commit

Table 8.6.
Correlation of Professional Practices and School Culture Factors with
Students' Mathematics Achievement Gains

		Correlation with Students' Mathematics Achievement Gains	Probability
Professional practices	Standards-based instructional practices	0.316	0.03
	Standards-based performance practices	0.204	0.17
	Homework practices	−0.389	0.01
School culture	Facilitative leadership	0.108	−0.53
	Learning community	0.325	0.05
	Teamwork	0.309	0.07
	Shared vision	0.217	0.20

to the particular activities noted in the item. Thus a high score on the *homework practices* scale indicates that teachers at the school reported giving considerable attention in their homework assignments to: working on demo projects or proofs; explaining reasoning or thinking; showing required steps; and taking a quiz or test on homework-assigned material. The results of this analysis support the assertion that these activities are better suited to the classroom than to homework simply because teachers are available during class time to assist when students get into trouble with assigned tasks. Further, the guidance of the teacher in carrying out these tasks may be critical for students to retain knowledge and information.

School Culture

Table 8.6 presents the correlations of the four scales of the SCQS with students' mathematics achievement gains. We found correlations in the thirties between students' mathematics achievement gains and the learning community (p < .056) and teamwork scales (p < .067) of the SCQS, together with lesser correlations for facilitative leadership and shared vision scales. Model 3 in Table 8.5 presents the results of a regression analysis with students' mathematics achievement gains as the dependent variable and the learning community, teamwork, and vision scales as the independent variables. Model 3 indicates that school-level scores on these three scales explained approximately 21 percent of the variability in school-level students' mathematics achievement gains (p < .015). Each of the school culture scales made statistically significant contributions in terms of the total percentage of students' mathematics achievement gains explained by the model. The addition of the facilitative leadership scale did not increase the variance explained by the model and thus was not included in our final model.

Combined School Culture and Instructional Practices Variables

When we think about instructional practices as supported by the school culture or climate, it makes sense to consider both sets of variables together, as we do in the last of our analyses. Model 4 presents the results of a regression analysis with students' mathematics achievement gains as the dependent variable and the already identified-as-significant learning community, shared vision, and teamwork scales from the SCQS, and the *standards-based instructional practices* and *homework practices* scales from the instructional practices group as independent variables. Examination of Table 8.5 reveals that 26.6 percent of the variability in school-level students' mathematics achievement gains is explained by the combined school culture and instructional practices variables (p < .01).

It is informative to compare the 26.6 percent of the variance in students' mathematics achievement gains scores explained by this combination of individually significant variables with the 25 percent of the variance explained by the learning community, teamwork, and shared vision scales of the school culture (SCQS) measure. It is obvious that there is a large proportion of common variance among these variables. This observation raises the chicken-and-egg controversy. Does school culture as defined by the SCQS contribute to instructional practices or do good instructional practices lead to a strong school culture? We recognize that, in the absence of repeated measures of the educational practices data and the school culture data, we cannot be sure about the direction of causal activity. It seems probable that this is not an either/or question, but that there is causal activity in both directions. In future studies we plan to attempt to survey faculty annually to begin to untangle this question. While our research does not definitively answer that question, we can say that both must be present in schools for students to make the kinds of achievement gains that are now required by recent federal policies as embodied in the No Child Left Behind Act.

CONCLUSION

In conclusion, our analyses suggest that when experienced teachers are provided sustained professional development emphasizing a standards-based approach to authentic instruction, especially at the school and classroom levels, student achievement is enhanced. Policies to support this kind of emphasis are critical to assure the kinds of outcomes that are necessary if all children are to learn to high standards so that, indeed, none will be left behind. While commitment to professional development at the school level to sustain reform must be present throughout the system, the perspective of district and school site administrators is critically important in defining the approach and emphasis given to professional development and in fostering communities of learners who benefit from a nurturing school culture. In addition, professional development at the school level that takes place in an atmosphere of trust and support, uses teaching case material, relies upon standards-based materials, and employs mathematics and science specialists as supports is likely to enhance teachers' classroom practices and result in student achievement gains, as the analyses in this chapter show.

9

What Have We Learned?
A Summary of Key Findings

Understanding the factors that enhance the impact of systems reform is critically important for many reasons. Perhaps the most important of these is that equitable reform of urban education is necessary to insure both a just and a prosperous society. Large urban districts educate 25 percent of all school-age students, 35 percent of all poor students, 30 percent of all English-language learners, and nearly 50 percent of all minority children (Hewson, 1998). We will fail utterly as a society if we do not work for equity in access to challenging materials in mathematics and science and opportunity to achieve to high standards for all children.

The major goals of the research study that has been the focus of this volume were: (1) to assess the impact of the reforms undertaken through the National Science Foundation's Urban Systemic Initiative by modeling relationships between latent variables (drivers) and sets of indicator variables (outcomes); (2) to determine how reforms in mathematics and science curriculum and instruction affect teacher practices and student outcomes at the classroom level; and (3) to investigate the roles of leadership, resources, and policies related to systemic reform that foster or inhibit student achievement outcomes and outcome differences. To achieve these goals, our research was organized to undertake three studies conducted simultaneously: (1) the Study of the Enacted

Curriculum; (2) the Policy Study; and (3) the Mathematics and Science Attainment Study. In this book's previous chapters we unpacked the results of these interrelated studies to show that effective systems reform requires the full-scale involvement and commitment of all individuals at all levels, including school district administrative leadership; teachers, principals, and community members at the level of the school; and ultimately, the teaching and learning that take place in the crucible of the classroom.

In this chapter we attempt to pull together a discussion of the most critical findings from the three years of research in the four cities and their districts. Overall, the research reported in this volume contributes to an understanding of how and under what circumstances systems reform is best put into practice at district, school, and classroom levels. This chapter is organized to report on our key findings at three levels of influence: the district, the school, and the classroom. It should be mentioned that the national policy environment is also critical, though it went unexamined systematically in our work. However, without national standards governing student learning and achievement in mathematics and science, and absent the resources and support of the National Science Foundation, systems reform is unlikely to occur.

WHAT MATTERS AT THE DISTRICT LEVEL

First, at the district level, we learned that when top-ranking administrators, including the superintendent or district CEO, adopt the objectives of the reform as the centerpiece for districtwide staff development and curriculum reform, they are able to affect positively student achievement over the period of reform. These findings are extremely important and show plainly how critical it is to have a coherent policy at the district level that pervades the full fabric of the district's mission, vision, and resource allocation efforts. Furthermore, even the largest districts can achieve coherence. Although our data point to Memphis as a district successfully engaged in systems reform, our data also clearly show that Chicago, the nation's third-largest district, was also highly successful in achieving the goals of the reform.

Findings from the Policy Study indicate that both district and school administrators viewed the provision of professional development opportunities as a primary focus for reform implementation (see chapter 2). District administrators varied, however, in their commitment to USI reforms and in the subsequent organization of reform implementation. Districts that saw the USI reform as the centerpiece for curriculum change were more effective in influencing implementation of USI reforms than districts with less optimal approaches.

Both Memphis's and Chicago's district chiefs embraced the USI reforms, embedded district staff development and curricular strategies in the implementation of the USI initiative, allowed little or no discretion in course taking, with the expectation that all students were to achieve in challenging coursework, and fully accepted the ideology that all students can achieve to high standards. These two USI sites made exemplary changes in the level of expectations and performance of teachers and other staff participating in our study, ultimately leading to both increased student achievement and dramatic decreases in the achievement gap. In El Paso, although the Collaborative efforts were exemplary in many respects—especially because the Collaborative successfully marshaled the support and involvement of a broad range of community constituencies—reforms were stymied by a deep commitment at the school and classroom level to enhancing student performance on the state-mandated TAAS. This commitment unfortunately was interpreted by a large number of teachers as requiring an approach to instruction akin to "drill and kill," and antithetical to the constructivist strategies that constituted the fabric of the NSF reforms.

When USI reforms are not the central focus or organizing principle for the district's mission or when high-stakes testing is perceived as assessing skills incongruent with hands-on, problem-solving approaches to student learning, student achievement is less likely to be affected. In the case of El Paso, for example, although virtually all administrators of the El Paso Collaborative (based at the University of Texas at El Paso) were dedicated to the USI reform agenda and had indeed engaged the community, many administrators in the three school districts served were uncomfortable with the governance structure, arguing that funds should flow directly to the districts. In addition, both Miami-Dade and El Paso were subject to rigorous statewide accountability systems that many district leaders and teachers saw as running counter to the goals and strategies of USI reforms. Finally, when districts are organized to insure adoption of reform throughout the system, higher levels of implementation of the reform are consistently maintained.

We also found variation at the district level in approaches to implementing the NSF reform agenda (see chapter 2). For example, some districts used professional development carried out in schools linked together in feeder pattern arrangements while other districts took a systemwide approach, especially if the district in question (Memphis, for example) was relatively small in size. In addition, districts varied in the manner in which stakeholders in the community were engaged with reform. In Chicago, for example, stakeholder involvement centered on the participation of museums and businesses in implementing mathematics and science reforms.

WHAT MATTERS AT THE SCHOOL LEVEL

At the level of the school, our results show the importance of the principal in promoting and maintaining a school culture that emphasizes teachers working and learning together as a community (see chapter 3). Our findings also show that while USI reform efforts have positive impacts on student achievement, more sustained, in-depth professional development programs targeted to specific classroom teacher needs may be necessary to significantly change teaching practices across the board. Professional development programs must simultaneously build the capacity for improvement as well as develop a school culture that nurtures a learning community in classrooms, schools, and school communities.

Teachers in schools undertaking high or even moderately high levels of reform reported spending between six and fifteen clock hours in content area professional development, compared with teachers in low-reform schools who reported less than six clock hours during a given calendar year. Teachers in high-reform schools reported using standards-based practices they had learned during the course of their professional development with greater frequency than teachers in moderate- or low-reform schools. When considering the specific nature of professional development activities, we found that mathematics professional development experiences that provided in-depth study of mathematics content and methods of teaching are significantly related to student mathematics test score gains over the four-year period of the USI reforms (.30). It is clear that teachers must be well grounded in mathematics and science content to be effective in their delivery of classroom instruction.

At the level of the school we also found several additional factors that determined the effectiveness of the implementation of USI reforms, leading in turn to increases in student achievement over the course of the reform. First, those schools judged by their respective districts to be high implementers of the reforms had relatively larger numbers of teachers participating in staff development related to USI reforms. In addition, staff development was carried out at the classroom level in these schools and was sustained over the duration of the reform. Further, teachers in high-implementation schools took more advanced mathematics courses and more mathematics and science methods courses than teachers in moderate- and low-reform schools. Moreover, these schools also were characterized as fostering communities of learners and providing opportunities for enhancing teaching skills related to USI reforms. In addition, schools with actively participating community-based stakeholders, including parents, community-based groups, and businesses, were most likely to have positive student achievement gains during the course of the reforms (see chapter 4). Finally, one site, El Paso, elected to move to block scheduling system-wide in an effort—in theory—to allow more time for teachers to develop student understanding by engaging them in challenging activities. However, this practice did not result in enhanced learning experiences and appeared to undermine student engagement in mathematics and science (see chapter 6). This finding raises concerns about teacher preparation for this type of classroom environment. Block scheduling is intended to provide teachers and students the opportunity to engage intensively in one or more subject matter areas, with increased student learning and understanding of the subject matter in question as the expected outcome. However, our research found that when schools used block scheduling teachers were more likely to use worksheets followed by extended periods of independent seat work. Clearly, districts using block scheduling must provide appropriate professional development and adequate materials in line with the goals of the reform to overcome the negative effects that we repeatedly observed.

WHAT MATTERS AT THE CLASSROOM LEVEL

The research findings show that teachers vary with respect to the amounts and types of staff development they have received; their capacity to engage in classroom instruction congruent with standards-based practices; and the impact at the classroom level on student achievement outcomes over the course of the reforms (see chapters 4 and 7). In addition, students vary systematically in their assessments of their teachers' classroom practices (see chapter 5). Further, these factors are significantly related to student achievement outcomes. Our observations as well as teacher and student survey results indicate that relatively small percentages of teachers are consistently engaged in implementing standards-based classroom practices.

Our analysis of the Study of the Enacted Curriculum's Survey of Classroom Practices (SCP) for teachers showed that teaching practices across each of the sites are more similar than they are different, with teacher responses on only thirty-one mathematics items and twenty-six science items (of 147) showing statistically significant outcomes by site (see chapter 5). However, distinctive patterns of classroom practices *did* emerge by site. Memphis and Chicago teachers evidenced a more standards-based orientation to instruction than teachers in either El Paso or Miami-Dade. That is, teachers in Memphis and Chicago reported greater impact from professional development on a number of classroom practices in contrast to teachers in El Paso or Miami-Dade who indicated that they either did not participate or that participation had little to no impact.

Additionally, analysis of researcher observations of mathematics and science lessons found that in 82 percent of all classrooms teachers engaged in what we term predominately teacher-centered or didactic teaching practices as opposed to standards-based approaches. In summary, findings from the teacher and student versions of the SCP and classroom observation analyses reveal that classroom practices across the four sites are more similar than not, and are more traditional or didactic than standards-based. However, there are intriguing cross-site variations favoring Memphis and Chicago.

We also found through the use of the Experience Sampling Method that higher and lower student engagement levels were associated with vastly different classroom practices (see chapter 6). Group work was a particularly important factor in this regard. Students responding to our ESM surveys argued that group work encouraged them to explore problems on their own and to work together with peers. It also provided a range of strategies for understanding problems posed in mathematics and science classes, ingredients for higher levels of engagement in classrooms. We also learned that group work introduced students to new sources of information, an outcome less likely to occur in the context of other class configurations. According to our student participants, group work allowed them to work out problems and come upon insights through discussions with their peers. It is clear that, based on these findings, professional development activities should stress the effective use of student group work, problem-solving activities, and active engagement with the curriculum.

Findings from the Mathematics and Science Attainment Study reveal that NSF's six-driver model explains the association among aspects of standards-based instruction, policy alignment, and resources related to professional development as 16.6 percent of the variability in the mathematics achievement components (see chapter 7). In Miami-Dade direct effects on student mathematics achievement gains were related to principals' views of the impact of professional development (.22), equity (−.25), time in professional development (.11), and professional development involving standards-based activities (.15). The total effect of instructional influences on achievement was positive (.10), while the total effect of stakeholders/community on achievement was negative (−.09). A comparable model, hypothesizing a seventh driver—school culture—explained 19.9 percent of the variance associated with gains in student achievement.

When measuring the achievement gap between USI schools and non-USI schools (as reported in chapter 8), we found our studied USI schools were performing at a level that was more than one-third of a standard deviation below our non-USI schools at the baseline. However, over the four years of the USI reforms, the USI schools closed that gap to just over one-tenth of a standard deviation. Our repeated measures analysis of variance yielded no main effect of time and a borderline main effect for group (in other words, studied USI versus non-USI) $F = 3.56$ (1220), $p = .10$. Further analyses revealed that differences between groups in 1996 and 1997 were significant ($p < .05$), while the differences between the groups in 1998 and 1999 were not, indicating movement toward closing the achievement gap.

In summary, those USI sites that aligned the district mission most closely with staff development, curriculum reform, and the belief system of the administrative leadership—one congruent with the value structure undergirding the purposes of systemic reform—fostered both professional growth and development among teachers of mathematics and science. This occurred in a manner that suggests the applicability of the NSF driver model. In other words, for reform to be successfully carried out systemwide, it is crucial to have:

- Programs designed to be strongly student-achievement-outcome-oriented
- Programs with explicit emphasis on resource convergence
- Partnerships that entail more than the provision of resources
- Advocacy at all levels of the system for program goals
- Standards-driven curriculum, instruction, and assessment practices
- A strong, vital school culture supporting and sustaining teachers' professional development and student achievement.

At the level of the school and classroom, the principal's leadership in channeling resources and reinforcing district goals congruent with systemic reform is vitally important, as documented in our study of the seventh driver, school culture. Most critical in the chain of interrelated factors leading to increased achievement and reduction of the gap are events that occur at the classroom level. Here we see that professional development is a preferred avenue to enhanced instruction according to teachers participating in our research, especially when it emphasizes both content and methods and occurs at the level of the classroom. The underserved students in our nation's cities cannot afford to have less than the best instruction available to them. This study demonstrates that even in the most challenged schools and classrooms students can achieve to the highest standards when the will, belief system, resources, leadership, and commitment are in place.

Appendix A: Instrumentation

This appendix contains the instruments designed or modified by the USI research staff during the course of the evaluation research grant from the National Science Foundation. Because the project was designed both to determine the impact of USI reform in four cities and to test the NSF six-driver model for reform, it was necessary to create instruments that would elicit information to meet both research goals. The instruments were designed to measure aspects of NSF's six drivers (see chapter 1, Table 1.2).

The appendix is organized into three sections corresponding to the three studies that constitute the scope and focus of our research. The three studies are: the Mathematics and Science Attainment Study; the Study of the Enacted Curriculum; and the Policy Study.

Mathematics and Science Attainment Study

The Mathematics and Science Attainment Study was geared toward understanding the types of instructional practices in mathematics and science classrooms and how these practices impact student achievement. Through classroom observations and teacher and student surveys of instructional practices (from the Study of the Enacted Curriculum), we were able to document the types of instructional practices being used during the USI reform. The student engagement study was conducted to determine which types of instructional practices kept students engaged during a lesson, and the School Culture Quality Survey sought to document the type of school culture present and whether or not it was supportive for reform activities. A brief discussion of each of these data collection methods is presented below along, with the instrument used to gauge it.

Classroom Observation Protocol.   In the fall of 1999 the research team designed the classroom observation protocol to collect information related to Driver 1, standards-based instruction. The process began by identifying the categories of standards-based practices for observation, using standards developed by the National Council of Teachers of Mathematics (NCTM) and the National Academy of Sciences as guides. The standards developed by the NCTM and the National Academy of Sciences emphasize high-quality mathematics and science education "for all students." They are also tied to the notion of the importance of resources to the enhancement of classroom instruction. Students cannot be taught with equity and fairness unless schools and classrooms are equipped with the necessary technological support, in addition to instruction that is standards-based and carried out by teachers who have appropriate certification and professional development experience.

Based upon that identification, the research team organized the protocol into three parts to collect data on instructional practices occurring in the classrooms of participating teachers. The first part included information about the class such as the number of students and the ethnicity and gender of the students and teacher. The second part identified the instructional materials, including technology, used during the observed lesson, as well as how the class was arranged. Finally, classroom events, such as teacher and/or student conversations, interactions, and activities were noted.

Classroom Observation Coding Checklist.   The classroom observation checklist instrument was used to code the classroom observations. The checklist consisted of mathematics and science standards as well as components of other instruments designed to identify five mathematics and science standards-based practices (NCTM, 2000; NRC, 1996):

- Communication is a way of sharing ideas and clarifying understanding.
- Problem solving is engaging in a task for which the solution is not known in advance. (Problem solving and inquiry were used interchangeably in this study.)
- Representation applies to processes and products that are observable externally as well as to those that occur "internally" in the minds of people, such as diagrams, graphical displays, and symbolic expressions.

Checklist of Observed Standards-Based Practices

1. Observer's Name _____   8. Teacher Gender _____
2. Date and Time _____   9. Teacher Ethnicity _____
3. School Name _____   10. Teacher Age (approx) _____
4. Teacher Name _____   11. Class Size _____
5. Grade Level _____   12. Ethnicity of Students (#'s) _____
6. Subject _____   13. Gender of Students (#'s) _____
7. Class Academic Level
   (if applicable) _____

Please indicate the emphasis placed on each of the statements: 0 = None, 1 = Minimal, 2 = Moderate, 3 = Major

Communication

0 1 2 3   Modeling appropriate use of mathematical and/or scientific language to communicate mathematical and/or scientific concepts. (Look for effective communication techniques, oral, written, and visual forms).

0 1 2 3   Providing students with the opportunity to communicate mathematical and/or scientific ideas using oral, written and/or visual forms (e.g., students explaining the process by which they reached a solution to a mathematical or scientific problem).

0 1 2 3   Guiding student communication of mathematical and/or scientific ideas (e.g., telling a student, "That wasn't completely clear, could you rephrase that?").

Problem Solving

0 1 2 3   Modeling the use of problem-solving strategies above and beyond presenting a specific technique or procedure (e.g., presenting a variety of strategies that can be used to solve a problem and encouraging students to experiment).

0 1 2 3   Providing students with the opportunity to work alone or in groups to discover solutions to problems.

0 1 2 3   Guiding students in exploring ways to solve a problem and/or in discovering the answers to questions (e.g., teacher asks questions to stimulate student thinking; asks students to share the different approaches they may have used).

Representation

0 1 2 3   Modeling the use of mathematical and scientific representations (e.g., using visual and/or tactile representations (graphs, tables, and models) of mathematical and/or scientific concepts to help explain these concepts to students).

0 1 2 3   Providing students with the opportunity to use mathematical and/or scientific representations individually and/or in groups to solve problems.

0 1 2 3   Guiding students in their use of mathematical and/or scientific representations to solve problems.

Reasoning

0 1 2 3   Modeling orally the mathematical and/or scientific reasoning processes (e.g., explaining the logic behind each step in a procedure or formula; explaining the reason why situations occur utilizing knowledge the students have been taught).

0 1 2 3   Providing students with the opportunity to develop reasoning skills (e.g., asking students to explain the procedure they used to solve a problem, asking students "why" questions for which the student has not been taught the answer).

0 1 2 3   Guiding students in use of reasoning (e.g., when students explain the procedures they used to solve a problem, teachers question further: "Why does _____ in this situation?").

Connections

0 1 2 3   Presenting and explaining one or more of the following:
1. The ways mathematical and science concepts covered in class can be and are used to complete real-world tasks.
2. The connections between various topics within mathematics and/or science and/or between mathematics and/or science and other disciplines.
3. The connections between the current lesson and mathematical and/or scientific concepts previously covered.

0 1 2 3   Providing students with the opportunity to explore, present, and/or discuss one or more of the following:
1. The ways mathematical and science concepts covered in class can be and are used to complete real-world tasks.
2. The connections between various topics within mathematics and/or science and/or between mathematics and/or science and other disciplines.
3. The connections between the current lesson and mathematical and/or scientific concepts previously covered.

0 1 2 3   Guiding student exploration and discussion of one or more of the following:
1. The ways mathematical and science concepts covered in class can be and are used to complete real-world tasks.
2. The connections between various topics within mathematics and/or science and/or between mathematics and/or science and other disciplines.
3. The connections between the current lesson and mathematical and/or scientific concepts previously covered.

0 1 2 3   Classroom displays depict the ways mathematical and/or science concepts covered in class are applied to real-world tasks.

- Reasoning involves developing ideas, exploring phenomena, justifying results, and using mathematical or scientific conjectures in all content areas.
- Connections are the interplay among mathematical or scientific topics, in contexts that relate those topics to other subject areas, and in students' own interests and experiences.

The five mathematics and science standards were organized to include three parallel statements describing teacher actions on a 0 to 3 scale: the first referred to whether the teacher modeled the particular standard; the second to whether the teacher provided students with the opportunity to engage in the process; and the third to whether the teacher provided appropriate guidance to students. This instrument identifies which standards the teachers were using as part of their instructional activities in the classrooms.

Authentic Instructional Practices Coding Matrix

The Authentic Instructional Practices Coding Matrix was modified by the research team to analyze classroom observations using two major sources: Authentic Instruction Classroom Observation Form (D'Agostino, 1996) and the Secondary Teacher Analysis Matrix (Simmons, Emory, Carter, and Coker, 1999). Using these sources as a guide, the Authentic Instructional Practices Coding Matrix is composed of four categories: classroom communication, social support, student engagement, and lesson coherence. The findings from the analysis identify whether teaching practices are teacher-centered (didactic), subject-centered (conceptual), or student-centered (constructivist).

Student Engagement Study

This study assesses the impact of classroom activities on high school engagement in ongoing mathematics and science teaching and learning activities using a methodological approach called experience sampling method (ESM) developed by Larson and Csikszentmihalyi (1983). Two instruments were used to determine what students are doing, thinking, and feeling throughout the course of a class: the student engagement survey and the student focus group. In addition, classroom observations were used to contextualize the student engagement survey, using the classroom observation protocol presented above.

Authentic Instructional Practices Matrix

Classroom Processes	Teacher-Centered (Didactic)	Subject-Centered (Conceptual)	Student-Centered (Constructivist)
Communication	The teacher is "telling." Content tends to be descriptive with little emphasis on explanation. There is no probing of student responses and no discussion of the lesson material. The teacher may ask questions that ask students to recall facts. There is very little communication between students or from students to teacher.	The teacher explains and analyzes the content, emphasizing procedural knowledge with explanations, and conceptual understanding maybe included. The teacher probes for student responses, but there is little conversation occurring among students— most dialogue is directed through the teacher.	Students are encouraged to converse among themselves about the lesson material. The teacher may or may not engage in the conversations. The teacher and students negotiate understanding of key ideas based on students' ideas and content. Investigations dominate conent. There is evidence that the purpose of sharing ideas is to arrive at a deeper understanding. Conceptual content and connections are embedded into the design, implementation, analysis, and report of investigations.
Problem Solving and Investigation	No explicit mention of "how we know." Problem solving and investigation are presented separately as rote procedure. The teacher may describe how to solve problems and students are expected to use those descriptions to solve similar problems. Note: "how we know" refers to how the conclusions were drawn.	The teacher describes "how we know" to the students by integrating the mathematical or scientific process with the concepts. The teacher explains the rationale of problem-solving steps and students are expected to use that rationale to solve similar problems.	Students, with or without teacher's guidance, reconstruct how evidence has been used to formulate and evaluate mathematical or scientific ideas. Students make and test conjectures (hypotheses) to solve problems.

228

Authentic Instructional Practices Matrix
(continued)

Classroom Processes	Teacher-Centered (Didactic)	Subject-Centered (Conceptual)	Student-Centered (Constructivist)
Representation	The teacher states the concepts so that the limits or exceptions within content are not presented. Most statements are absolutes without qualifiers or are oversimplified. Representations are not used to develop concepts.	The teacher demonstrates the concepts using a specific representation. representations are used to enhance understanding. Alternate representation of ideas may not be encouraged or discussed.	The teacher and students generate multiple ways of representing or interpreting observations and events. They identify limits and exceptions that exist within each representation.
Reasoning	The teacher requires students to recall factual information.	The teacher asks students to justify or explain ideas using information that has been taught.	The teacher encourages and provides opportunities for students to make, test, and justify conjectures/hypotheses.
Connections	Lesson topic and activities have no clear connections (examples or interconnections) to: a. real-world events, such as experiences with media or their community b. related ideas within the subject c. related ideas between different disciplines	The teacher tries to connect the topic of the lesson indirectly or directly to: a. real-world events, such as experiences with media or their community b. related ideas within the subject c. related ideas between different disciplines	Connections are constructed by students with or without teacher's guidance to: a. real-world events, such as experiences with media or their community b. related ideas within the subject c. related ideas between different disciplines

229

Authentic Classroom Dynamics Matrix

TOPIC	ZERO	ONE	TWO	THREE
Social Support	The rapport between teacher and students is not good. The working relationship between teacher and students is rarely constructive. The over-all atmosphere of the classroom is negative.	Support is mixed. Teacher praises students occasionally. At other times student effort goes unnoticed. Students are not encouraged to support one another.	Support is usually positive. Teacher-student rapport is good. There is some evidence of high expectations for learning and trying hard. Teacher focuses on student successes and does not dwell on failures.	A strong friendship and mutual trust exists between teacher and students. The atmosphere clearly supports student effort. Lowest achieving students receive support from all.
Student Engage-ment	Students appear to be inattentive. They may look as though they are bored or preoccupied with thoughts unrelated to the task at hand. One or a few students may be disruptive.	Students appear to be occasionally on-task. For those that are on-task, however, they seem to be rather lethargic and/or not trying very hard.	Students for most of the time are on-task pursuing the substance of the lesson. Students have, however, occasional lapses in concentration. Students, with few exceptions, are on-task.	All but one or two students are deeply engaged in the lesson (paying attention, clearly interested in learning the material, concentrating) for all but a few short instances of the lesson.
Coherence of Lesson	Material is presented in superficial fragments with very little connection between parts.	Some **activities** focus on significant topics, but other key concepts or ideas are not appropriately covered. The lesson activities are not well connected to overarching concepts or ideas. (Activities for the sake of doing activities.)	Some key **concepts or ideas** are covered in depth. There are activities that focus on significant topics that are key to the whole lesson content, but coverage is uneven and other key concepts or ideas are not appropriately covered. There are some good parts of the lesson, but there are parts that are missing or do not appropriately support concept development.	Key concepts/ideas are covered in depth. The lesson content is presented as a whole and is structured in a way that allows for the sequencing and structuring of a complex topic. Each topic appears to build on another in an effort to foster deeper student under-standing.

230

Students were asked to complete the engagement survey when they were signaled between two and four times, depending on the length of the class period, using a vibrating hand-held device (which was placed on their desk). The same questions were asked with each signal; thus, the instrument provides time-series data on student engagement moment-to-moment. Through the use of the instruments mentioned above, the study provides student-level information on Driver 1, standards-based instruction and how instruction relates to student engagement. Driver 4, mobilization of stakeholders, is discussed during the focus group.

FIRST SIGNAL

1. When you were signaled the first time today, what was the main thing that you were doing?
2. What else were you doing?
3. What was on your mind?

Please fill in the bubbles: Strongly Disagree = SD, Disagree = D, Agree = A, Strongly Agree = SA

When you were signaled the first time today,	SD	D	A	SA
I was paying attention to class.............	O	O	O	O
I did not feel like listening...................	O	O	O	O
My motivation level was high............	O	O	O	O
I was bored	O	O	O	O
I was enjoying class..............................	O	O	O	O
I was focused more on class than........ anything else..	O	O	O	O
I wished the class would end soon	O	O	O	O
I was completely into class..................	O	O	O	O

Please fill in the bubbles.
When you were signaled the first time today, was the main thing you were doing more like:

  Work O Play O Both O Neither O

What was being taught was:

  Very easy O Easy O About right O Difficult O
  Very difficult O Can't tell O

What was being taught was something that I already knew:

  Yes O No O Can't tell O

Please fill in the bubbles.

When you were signaled the first time today, what was being taught was important for:

my everyday life	Yes ○	No ○
going to college	Yes ○	No ○
my future job	Yes ○	No ○
future tests	Yes ○	No ○

If you felt that class *at the time of signal* was important for tests, please tell us for which tests it was important. Fill in *all that apply*.

Class quiz ○   Midterm/Final ○   SAT or ACT ○

State Assessment Tests (e.g., TAAS) ○

Class was unrelated to test ○

Please fill in the bubbles.

When you were signaled the third time today, you were feeling (Fill in *all that apply*):

Happy	○	Confused	○	Active	○	Having fun	○
Nervous	○	Intimidated	○	Sad	○	Cooperative	○
Relaxed	○	Worried	○	Angry	○	Confident	○
Competitive	○	Frustrated	○	Busy	○	Sleepy	○

Were you talking with anyone? No one ○   Classmate ○   Teacher ○

Was it about the class?   Yes ○   No ○

Comments—if any:

Student Focus Group Interviews

FORMAT: 8–10 students in a group; 2 moderators—one to record and one to facilitate; tape recorder to capture narrative.

1. What math and/or science classes are you taking this year?
2. Are you and your friends in the same math and/or science classes?
3. What kinds of extracurricular activities (clubs, sports, etc.) are you involved in at school?
4. What kinds of activities, organizations, etc., are you involved with in your neighborhood, in the community? What do you

do with these groups? Do you think your work with them helps you in school? How?
5. What do you intend to do after high school?
6. Do your future plans involve using math, science, or technology?

School Culture Quality Survey

The purposes of this survey were to determine whether the quality of professional community as measured by the School Culture Quality Survey (SCQS) (Katzenmeyer, 1999) was positively related to student achievement in mathematics and whether a school culture driver added to the NSF driver model might enhance our ability to predict the values of the outcome drivers. The survey, as part of the Mathematics and Science Attainment Study, sought to determine whether or not significant relationships exist between measures of teachers' professional community (school culture) and student's achievement in mathematics.

Based on ideas and concepts from the quality movement and the work of Peter Senge, the SCQS conceives the quality of professional community to include the following factors: shared vision, facilitative leadership, teamwork, and learning community. Beginning in 1996, the SCQS was developed to assess the extent to which a school professional community characterized by these four elements is present in schools.

Information about the complete School Culture Quality Survey is available at http://anchin.coedu.usf.edu.

Study of the Enacted Curriculum

The study of the enacted curriculum consists of three surveys for teachers and students in mathematics, with three parallel surveys for science. Mathematics and science teachers complete two surveys: a survey of instructional content and a survey of classroom practices. The survey of instructional content is not reported in this book. Students complete the student survey of classroom practices for either science or mathematics. Together, these surveys elicit teacher and student responses on what is occurring in mathematics and science classrooms in terms of content, instruction, professional development, technology, and homework.

Information from these surveys will not only inform the Study of the Enacted Curriculum, but the Mathematics and Science Attainment Study as well. These self-reported data can be compared with the classroom observation data and lend more information to Driver 1, standards-based instruction. Additionally, the survey provides information on professional development and other resources (Driver 3, Unified Application of Resources). The surveys can also determine if there are differences in mathematics and science teaching that are related to state policy initiatives and state standards (Driver 2, Unified Set of Policies).

The Study of the Enacted Curriculum was a subcontracted study through the Council of Chief State School Officers. The instruments had been in use since 1999, and collaboration with the USI study presented an opportunity to extend the research on classroom curriculum and practices by including the schools in our sample.

Information about the complete Study of the Enacted Curriculum is available at http://www.ccsso.org/projects/projects.html.

Policy Study

The policy study is composed of data from school and district administrators, teachers (through the use of focus groups, a questionnaire, and survey correspondence), and community stakeholders. The purpose of the policy study is to uncover the impact of state, district, and USI policies on the six drivers.

Principal Interview. The principal interview protocol consists of thirteen multipart items related to NSF's six-driver model. In addition to providing responses to questions asked during the interview, our participants were also asked to rate the level of importance in implementing reform of specific strategies such as "Identifying teachers' classroom instructional needs and providing them support" or "Evaluating the potential value of new instructional strategies and programs."

The principal interview questions were devised to provide data for each of the drivers in the NSF model. The interview protocol elicited information about individuals' roles as principal, their definition of standards-based instruction, their views of policies impacting mathematics and science, the impact of professional development, social equity issues impacting their schools, and the kinds of resources available, as well as the coordinated use of these resources. All together, the principal interview data can be used to support analysis for any of the drivers. It was also determined that the administrators would be asked to rate the importance of some of the topics on a zero to three scale, with zero being not at all important and three being extremely important. This rating scale has enabled the principal responses to be quantified for use in the study's six-driver model.

Principal Interview Protocol

General Context

Gather information about the principal:

Name:_____ Age:_____ Gender:_____

Race/Ethnicity:_____

How long have you been a principal?

How long have you been a principal at this school?

Have you held other administrative positions?

How many years have you taught?

What was the area?

What is your previous nonacademic work experience?

What is your highest level of education?

Specific Context

I would like to ask you about a dozen questions. In addition, I will ask you to use a rating scale on some:

0 = Not at all important to 3 = Extremely important

Q1: (Driver 7) How would you define your role as principal? *Probe*: How do you support instruction and learning in your school?

How would you rate the importance of:

1a) Identifying teacher needs and providing them support with regard to instruction?

0 = Not at all important to 3 = Extremely important

Please provide examples of your efforts in this area.

How would you rate the importance of:

1b) Making program adjustments in response to unmet student needs?

0 = Not at all important to 3 = Extremely important

Please provide examples of your efforts in this area.

Q2: (Driver 1) How would you define standards-based instruction?

2a) How is standards-based instruction implemented in the math and science classrooms at your school?

Please rate the impact of standards-based instruction on the curriculum

MATH   0 = Not at all important to 3 = Extremely important

SCIENCE   0 = Not at all important to 3 = Extremely important

Q3: (Multiple Drivers) Please detail changes that have occurred in your school since the implementation of the USI.

Please tell us the specific ways that the USI has affected the following in both math and science:

3a) Student Achievement

MATH   0 = Not at all important to 3 = Extremely important

SCIENCE   0 = Not at all important to 3 = Extremely important

Please provide examples.

3b) Professional Development
    MATH   0 = Not at all important to 3 = Extremely important
    SCIENCE   0 = Not at all important to 3 = Extremely important
    Please provide examples.

3c) Technology
    MATH   0 = Not at all important to 3 = Extremely important
    SCIENCE   0 = Not at all important to 3 = Extremely important
    Please provide examples.

3d) Social Organization (block class structuring, teacher prep time)
    MATH   0 = Not at all important to 3 = Extremely important
    SCIENCE   0 = Not at all important to 3 = Extremely important
    Please provide examples.

Q4: (Driver 2) Can you identify those policies (national, state, and district) regarding math and/or science instruction/achievement that have the biggest impact on your school?

    4a) What do you consider to be your role in implementing national and state policies at your school?

    4b) What are the policies that support or hinder the implementation of the USI in your school?

Q5: (Driver 6) A major policy emphasis in school reform is the reduction of social inequities. What social equity issues or problems are particularly troublesome at your school?

    5a) Do you feel that the USI has been useful in addressing such issues/problems? Why or why not?

Q6: (Driver 2) A lot of emphasis is placed on accountability and evaluation in reform. As a principal, what measures or processes do you use to assess teacher performance, either formally or informally?
Please rate its impact on the classroom.
    MATH   0 = Not at all important to 3 = Extremely important
    SCIENCE   0 = Not at all important to 3 = Extremely important

Q7: (Driver 3) During the last three years have you or your staff attended any professional development activities?

    7a) When during the year? What topics or programs were covered? Who sponsored the professional development?

    7b) What are the percentages of staff who attended? For staff who do not attend, how is information disseminated?

    7c) How many of your teachers are earning advanced degrees? Do you offer tuition reimbursement?

Q8: (Driver 3) How many out-of-field teachers are there in your school? What is the impact of the out-of-field teachers on meeting the USI goals?

Q9: (Driver 3) How often are meetings among staff at your school held to discuss instructional practices and other issues related to the USI? (Frequency, % of attendees, topics)

Please rate the impact of these meetings on the math and science programs at your school.

      MATH   0 = Not at all important to 3 = Extremely important
      SCIENCE 0 = Not at all important to 3 = Extremely important

Q10:   (Driver 3) Has participation in the USI enabled you to increase the amount of available resources (technology, instructional materials, funding, time)? Provide examples.

Q11:   (Driver 4) Thinking about community resources such as university partnerships, businesses, parents, etc., how has their participation in USI-related or other reform efforts impacted your school? (What are the other reform efforts?)

Please rate their participation as it impacts the science and/or math programs in your school.

      MATH   0 = Not at all important to 3 = Extremely important
      SCIENCE 0 = Not at all important to 3 = Extremely important

Q12:   (Driver 3) How are resources coordinated at your school and within the school district?

Please rate the impact of this coordination on the science and/or math programs in your school.

      MATH   0 = Not at all important to 3 = Extremely important
      SCIENCE 0 = Not at all important to 3 = Extremely important

Q13:   (Driver 5/6) How do you know if students are benefiting from your school's involvement in the USI?

     13a) Are there particular achievement markers of student achievement/accomplishment you consider important?

Please rate the impact these have on your school.

      MATH   0 = Not at all important to 3 = Extremely important
      SCIENCE 0 = Not at all important to 3 = Extremely important

District Interview. The district administrator interview protocol consists of fourteen multipart items related to NSF's six-driver model. In addition to providing responses to questions asked during the interview, participants were also asked to rate the level of importance in implementing specific reform strategies.

In addition to providing data to inform the complete policy study, responses from the interviews with district administrators provided alternative input for analysis of teacher professional development and interviews with principals.

Items for the district administrator protocol were identified and developed by the research team. Questions were developed to elicit feedback from district-level staff related to each of the NSF drivers.

To distinguish between curriculum specialists ("Curr") and other district-level administrators ("All"), a column labeled "Role" was added to the left of the questions. This distinction was made due to the content of a couple of questions, namely, questions two and nine. These questions required knowledge that curriculum specialists would possess, and not necessarily other district administrators.

USI District Interview Protocol

General Context
State the main purpose of our study and the goals of this interview.
Review the Informed Consent form.
Gather information about the district person:
Name: _____ Age: _____ Gender: _____ Race/Ethnicity:_____

- How long have you been working at the district level? How long have you been in this position?
- What other administrative positions have you held?
- How many years did you teach, and what was the subject area?
- What previous nonacademic work experience have you had?
- What is your highest level of education?

Role	Q #	Questions	Driver
All	1	Based upon your experiences during the past several years, what do you believe is the best way to improve mathematics and science education in your district?	1
	1a)	Currently, national efforts to improve education use the term reform. What do you believe these national reform initiatives are and how do they apply to your district?	
	1b)	The National Science Foundation, in the Urban Systemic Initiative, defined reform as standards-based reform that includes implicit constructivist principles. To what extent does the USI definition of reform support or conflict with your views?	
	1c)	(Driver 2, 3) How would you define your role as part of the Urban Systemic Initiative in your district? *Probe*: How do you support the reform agenda in schools?	

(continued)

USI District Interview Protocol (cont.)

Role	Q #	Questions	Driver
	1d)	How does the district identify teachers' instructional needs? Please provide examples of your efforts in this area.	
	1e)	How does the district support the identified instructional needs of teachers? How do you measure the effectiveness of these efforts?	
		1c) How does the district identify unmet student needs? Please provide examples of your efforts in this area.	
		1d) How does the district support these unmet student needs? How do you measure the effectiveness of these efforts?	
		1e) How does the district identify the needs of administrators? How does the district support the identified needs of administrators?	
Curr	2	How would you define standards-based instruction?	1
	2a)	How is standards-based instruction implemented in the math and science classrooms in your district?	
	2b)	How have you integrated standards-based instruction into the mathematics and science curriculum?	
	2c)	How do you determine the effectiveness of standards-based instruction?	
	2d)	How has the USI affected student achievement in mathematics?	
	2e)	How has the USI affected student achievement in science?	
	2f)	How has USI affected technology in mathematics?	
	2g)	How has USI affected technology in science?	
All	3	Please describe any major changes that have occurred in your USI plan since its implementation.	

(continued)

USI District Interview Protocol (cont.)

Role	Q #	Questions	Driver
	3a)	Why were these changes made?	
	3b)	What new or different organizational structures or patterns have been created as a result of the USI? Please provide examples.	
All	4	What policies (at the national, state, and district levels) regarding math and/or science instruction/achievement have the biggest impact on your district?	2
	4a)	What do you consider to be your role in implementing national and state policies in your district?	
	4b)	What policies support or hinder the implementation of the USI in your district?	
	4c)	What national, state, or local politics influenced decisions concerning your USI?	
	4d)	How have NSF's USI drivers influenced local policy? (Include a handout summary of drivers.)	
	4e)	Has the NSF USI influenced policy making in other areas?	
	4f)	What kind of working relationship do you feel the district has with the NSF in administering the USI?	
	4g)	Has that relationship changed during the course of the grant?	
	5	What social equity issues or problems are particularly troublesome in your district?	2
	5a)	What policies or procedures have been implemented to reduce social inequities?	
	5b)	How has the USI been useful in addressing such issues/problems?	
	6	What is the district's policy for evaluating school-based administrators?	3, 7
	6a)	What measures, criteria, or processes are used to assess administrative performance, formally and informally?	

(continued)

USI District Interview Protocol (cont.)

Role	Q #	Questions	Driver
	6b)	How do you determine the effectiveness of these evaluation procedures on student achievement in mathematics and science?	
	6c)	How are the procedures for evaluating district-based administrators different from school-based administrators?	
	6d)	How has district policy addressed improving the climate of the working environment at the district level, the school level, and the classroom level?	
	7	During the last three years have you or your staff attended any professional development activities?	3
	7a)	When during the year? What topics or programs were covered? Who sponsored the professional development programs?	
	7b)	What are the percentages of staff attended? Is attendance required? For staff that does not attend, how is information disseminated?	
	7c)	In what manner does the district support teachers and administrators to earn advanced degrees?	
	7d)	What process did you use in developing the agenda for professional development programs?	
	7e)	How is staff development designed, as a series of events or as part of sustained engagement?	
	7f)	What is the role of standards-based instruction and constructivist teaching practices in your professional development program? How was that role established?	
	7g)	How does staff development promote changes in attitudes and instrumentation?	
	8	What are the current policies regarding out-of-field teachers in your district?	3

(continued)

USI District Interview Protocol (cont.)

Role	Q #	Questions	Driver
	8a)	What steps has the district taken to reduce the number of out-of-field teachers in mathematics and science?	
	8b)	What steps have been taken to improve the mathematics and/or science content backgrounds of elementary teachers?	
Curr	9	How often are district meetings held to discuss the philosophical basis of instructional practices and other issues related to the USI? (Frequency, % of attendees, topics.)	3
	10	How has participation in the USI enabled you to increase the amount of available resources (technology, instructional materials, funding, time)? Provide examples.	3
	10a)	What policies were developed or implemented to handle the problem of having USI funds supplant rather than supplement district support for mathematics and science programs?	
	10b)	How were USI funds used to supplement district contributions? Please describe examples.	
	10c)	Were there instances where USI funds supplanted normal district support for math and science programs? Why did this happen?	
	11	What steps have you taken to insure the sustainability of the USI reform effort?	3
	12	Thinking about community resources such as university partnerships, businesses, parents, etc., how has their participation in USI-related or other reform efforts impacted your district? (What are some other reform efforts?)	4
	12a)	Please rate their participation as it impacts the science and/or math programs in your district.	
	13	How do you know if students are benefiting from your district's involvement in the USI?	5, 6
	13a)	What are the particular markers of student achievement/accomplishment for mathematics and/or science that you consider important?	

Teacher Focus Group. The focus groups were organized to inform the teachers about the research project, answer any questions, and obtain the opinions of the teachers about professional development opportunities they had experienced. The focus group facilitators asked the participants open-ended questions about which of their professional development experiences had provided the most useful or least useful information. Researchers also asked about what teachers wished to have changed about the professional development offered through the USI or district.

The teacher focus groups about professional development lent information about Driver 3 (Unified Application of Resources) in the six-driver model. One of the indicators for Driver 3 is professional development activities for teachers. In their responses to open-ended questions, teachers shared their experiences with professional development activities and identified the types that had the most impact on their instructional practices.

Protocol for Teacher Focus Group. Introduce the teachers to the NSF grant and ask if they have any questions.

Q1: We'd like you to reflect on your experiences with staff development over the last four years for mathematics and science.
Q2: Which professional development experiences were the most useful or helpful for you in your classroom instruction?
Q3: Which professional development experiences were the least useful or helpful for you in your classroom instruction?

Teacher Questionnaire. The teacher questionnaire was used in conjunction with the teacher focus group protocol to solicit individual teacher's points of view. The open-ended nature of the questionnaire allowed each teacher an opportunity to describe his or her personal views regarding professional development.

The teacher questionnaire informed NSF's six-driver model by providing information on several of the drivers from the teacher's perspective: Driver 1 (standards-based instruction), Driver 3 (unified application of resources), and Driver 4 (mobilization of stakeholders). The professional development questions (Driver 3) included in the questionnaire were used with the teacher focus group data to understand the teacher's professional development activities since the beginning of the USI reform efforts in each of the participating cities.

The items for the teacher questionnaire were developed by the research team to address how the USI reform has affected the teachers since the inception of the reform.

Teacher Questionnaire:

1. What has been your experience with the district's Urban Systemic Initiative? What are your impressions regarding the goals of the district's Urban Systemic Initiative?
2. Have you made any changes in your instructional strategies since the implementation of the district's Urban Systemic Initiative? If so, please describe these changes.
3. What types of assessment strategies (tests, oral reports, projects) do you use? Please give some specific examples. How did you choose the methods that you use?
4. What types of professional development activities (at your school or elsewhere) have you participated in through the district's Urban Systemic Initiative? Please comment on the content and format of these conferences/workshops.
5. Please comment on the effectiveness of these professional development activities. What aspects have you found useful or not useful? What types of professional development activities would you like to see in the future?
6. What instructional materials (if any) have been made available to you by the district's Urban Systemic Initiative? Please comment on the effectiveness of these materials and describe what would be effective in the future.
7. What type of input have you received regarding curriculum and instruction from the following:
   a) Parents?
   b) Principal?
   c) Other teachers?
   d) School board?
   e) Superintendent?
   f) University partnerships?
   g) Local businesses?
8. In what ways do you utilize community resources and/or school partners (individuals or organizations) in your classroom?

Teacher Survey Correspondence.   The correspondence topics were developed by the research team and administered throughout the school year to establish an ongoing communication with teachers concerning their views on the implementation of the USI. The topics addressed critical aspects of the NSF drivers: policies, student assessment, classroom assessment, technology, and curriculum decision making and sought to determine the impact of these on teaching practices. These topics were selected from the teacher questionnaire (see Teacher Questionnaire description) to gather more in-depth information on the topics from the teachers. The questions were open-ended and designed to provide rating data using a zero to three scale. The first question on each correspondence included three parts: Part 1 asked teachers to identify important aspects about each topic; Part 2 asked teachers to rate the topic's impact on instructional practices; and Part 3 asked teachers to describe the topic's use in their classroom. The first question was followed by three open-ended questions asking respondents to elaborate and evaluate the impact of each topic on their classroom practices. E-mail accounts were set up so that teachers could respond using the Internet. There were challenges to this method, so teachers faxed their responses to the research team instead.

Teacher Correspondence #1: January 2000

TOPIC 1: What mandates and policies (that is, the measures you are held responsible for implementing) most affect your classroom instruction in mathematics and science?

1. Please provide a brief description (exact title not required) of each policy and/or mandate.
   Indicate whether you perceive it to be a national, state, district, or school policy/mandate.
2. Indicate the extent of its effect on your instructional practices.
   0 = None   1 = Minimal   2 = Moderate   3 = Major
3. What is your professional evaluation of each policy and/or mandate with respect to its impact on student academic success?

Question #1: Policy	Question #2: (0–3)	Question #3: Evaluation of policy

Teacher Correspondence #2: February 2000

TOPIC #2: How does assessment affect your classroom instructional practices in mathematics and science?

Please provide the names of state, district, school, and classroom level assessment instruments (exams, tests, portfolios, etc.) for students in your school in general and your class in particular.
Indicate the extent of impact of each on your instructional practices.
0 = None   1 = Minimal   2 = Moderate   3 = Major

When and how do you prepare students for them? How much time does this preparation entail?

Question #1: Name or Description	Question #2: (0–3)	Question #3: Preparation Description

What particular issues concerning assessment has your faculty talked about and how did you resolve them?

As a school, what has been done to address the assessment needs of the student body?

Teacher Correspondence #3: March 2000

TOPIC #3: How have your assessment and grading procedures in mathematics and science been affected by the USI?

1. In column 1 please name the categories that you use for determining student grades in mathematics and science. These might include Tests, Quizzes, Projects, Presentations, Labs, Portfolios, Student Self- and Peer Evaluations, etc.
2. In column 2 please describe the role of each category in your assessment process; e.g. why it is included, how often it is used, when it is used, whether it "counts" every time, etc.
3. In column 3 indicate the effect of professional development on the inclusion of those categories:
   0 = None   1 = Minimal   2 = Moderate   3 = Major

Question #1: Name	Question #2: Category Description	Question #3: (0–3)

4. Please describe any professional development that affected your decisions to include particular categories in your assessment and grading process.
5. Please describe the processes by which you determine cumulative grades for particular time periods, including end of marking periods, semesters, and/or end of year.
6. Please describe any changes in your assessment and grading procedures in mathematics and science over the past five years.

For one math or science class, please provide a copy of your grade book pages from the beginning of the 1999–2000 school year through the end of the most recent grading period. *In the place of student names, which should be blocked out, please provide race and gender identifiers.*

Teacher Correspondence #4: April 2000

TOPIC #4: How does technology affect your classroom instructional practices in mathematics and science?

Please identify the forms and instruments of technology that you and your students use in the classroom. (Column 1)

Indicate the extent of impact of each on your instructional practices. (Column 2)

0 = None 1 = Minimal 2 = Moderate 3 = Major

When and how are they used? (Column 3)

Question #1: Name or Description	Question #2: (0–3)	Question #3: Description of Use

How available is technology to both teachers and students in your school?

Are there provisions in place to assure equal access to technology for all students and teachers? If so, how effective are those provisions?

How have your instructional practices changed as a result of technology?

Teacher Correspondence #5: May 2000

TOPIC #5: How does the curriculum decision-making process affect your classroom instructional practices in mathematics and science?

Please identify the various people that make decisions about what, when, and how you teach.

Indicate the extent of impact of each on your instructional practices.

0 = None 1 = Minimal 2 = Moderate 3 = Major

Describe the impact of each decision on your instructional practices.

Question #1: Name	Question #2: (0–3)	Question #3: Description of Impact

Describe your school's formal and informal decision-making processes for curriculum and instruction—that is, the ways in which you and your colleagues decide what to teach, how to teach, and when to teach it.

How would you evaluate the decision-making process for curriculum and instruction in your school? In your district?

Study of the Community Context

The instruments for the Study of the Community Context consisted of three protocols: one for school-level staff; the second for parent organizations, school governance, district staff, corporate, and community partners; and the last for parents of students. They were created for several purposes: to facilitate semi-structured interviews to gather information about the mobilization of stakeholders in each of the four sites; to gather information about stakeholders and partners at two high schools in each site; to help researchers understand the multiple definitions of "community" and "partners"; and to secure measures of importance and effectiveness of partnerships and stakeholder relationships.

Questions on the principal interview and district interview protocols were used in the design of the community study protocols in an attempt to collect similar information from multiple sources. The questions were adapted to gather more specific information about the roles and responsibilities of the respondents.

Protocols for Stakeholder Interviews
Interview Protocol #1: School-Level Staff
>(Partnership coordinators, school-to-work staff, counselors, volunteers, teachers)

```
Name: _____    Signature:_____
Position: _____
Gender: _____F          _____M
Race/Ethnicity:  White (non-Latino) _____      Latino _____
                 African American _____
                 Asian/Pacific Islander _____
                 Native American _____        Other _____
```

- How long have you been in (this organization)?
- What is your role/position in (this organization)?
- How did you come to occupy this role/position?
- Can you tell me what your position consists of (i.e., what are your responsibilities)?

Stakeholder Questions

1. A. In what ways do you think partnerships and stakeholder involvement in the school have changed since the recent USI reform (1994–2000)?

B. Can you rank the importance of stakeholder involvement to students' achievement and attainment on a scale of 0–3? Please explain your ranking.

0	1	2	3
Not important			Extremely Important

C. Can you rank the effectiveness of stakeholder involvement to students' achievement and attainment on a scale of 0–3? Please explain your ranking.

0	1	2	3
Not effective			Extremely effective

2. Definitions and perceptions of "community"
   A. How would you define the boundaries of the school's "community?" (Probe for specifics about neighborhoods or the community surrounding the school.)
   B. Would students and parents use the same definition(s)? How might they differ?
   C. How does the local community or neighborhood relate to the school (e.g., perceptions, activities, and involvement with either the school or with the partnerships related to the school)?
3. A. Can you identify any "partners" that engage with the school to assist the school in its efforts of reform? Please rank them in terms of their importance for improving students' math and science achievement and attainment (0—not important to 3—extremely important)

0	1	2	3
Not important			Extremely Important

B. Please rank the effectiveness of these "partners" in contributing to the improvement of math and science achievement and attainment (0—not effective to 3—extremely effective)

0	1	2	3
Not effective			Extremely effective

Interview Protocol #2: District-level Staff/Corporate and Community Partners

A. How long have you been involved with (this organization)?
B. How are you involved?

C. What prompted your involvement?

D. Do you hold a specific position within (this organization)?

E. Are there specific terms (length of participation) associated with your position?

2. Are you aware of the reform efforts at the school? (Probe for knowledge of USI. Use names of programs or activities funded by USI if respondent is not familiar with USI specifically.)

3. A. How would you define the boundaries of the school's "community"?

B. Would students and parents use the same definition(s)? How might they differ?

C. Do you think the math and science reform has affected the community (Probe: either the school's "community" or the community-at-large)? If yes, how?

4. A. Can you identify any "partners" that engage with the school to assist the school in its efforts of reform? Please rank them in terms of their importance for improving students' math and science achievement and attainment (0—not important to 3—extremely important)

B. Please rank the effectiveness of these "partners" in contributing to the improvement of math and science achievement and attainment (0—not effective to 3—extremely effective)

0	1	2	3
Not effective			Extremely effective

Interview Protocol #3: Parents of Students (Student Engagement Study)

1. What do you know about the Urban Systemic Initiative for math and science reform at your child's school? (Probe with names of specific programs or activities funded by USI.)

2. How much do you know about the reforms in math and science at your child's school?

3. Have you seen any changes in your child's math and science instruction since he or she started high school?

4. Is the homework different from the homework (schoolwork) that you did in high school? If so, how?

5. If you have older children, is the homework (schoolwork) different from the homework the older children did in high school? If so, how?

6. Do your child's classes seem "different" than classes were five years ago?

7. What does your child's school tell you about what is going on, specifically math and science instruction or reform? How do you get the information (flyers, brochures, calls, letters, other)?

8. How does your child's school help you better understand the math and science reforms? PROBE
Events like math fairs or science fairs
Workshops on homework assistance
Workshops on understanding the reform programs
Training for parents in the new instruction and course content
Family math or family science nights at the school
9. Have you attended any of these?
10. Do you think that these kinds of support need to be available to parents?
11. A. What do you know about the math and science course work needed for high school graduation or college?
    B. How important to you is your child taking algebra, geometry, and the science courses necessary for admission to college? Rank from 0—not important to 3—extremely important. Please explain your ranking.
12. A. What does your child's math and/or science teacher tell you about math and science instruction or reform?
    B. How do you get the information (conferences, meetings with the teacher, flyers, calls, letters, other)?
13. What classes are your children taking? (PROBE—IB, AP, etc.)
14. Do you help your child to choose his/her class schedule?
15. Please tell us your impression of your child's experience with math instruction. PROBE
What do you hear him/her saying about the class?
The class content?
Do you see your child using his/her math and science knowledge in activities or projects around the house?
16. Please tell us your impression of your child's experience with science instruction. PROBE
What do you hear him/her saying about the class?
The class content?
Do you see your child using his/her math and science knowledge in activities or projects around the house?
17. Testing—tell us what you think about the state testing being done in schools today (TCAP, TAAS, FCAT, CASE.)
Do you help your child prepare for these tests? If so, how?
Do you understand how these tests and scores are used to guide your child's school program?
18. What resources in your neighborhood are available to help your child improve/strengthen his or her math or science performance?

Appendix B: Survey of Classroom Practices

Table B.1.
Mathematics and Science Student and Teacher Surveys of Classroom Practice

| | Mathematics | | | | Science | | | |
| | Student** | | Teacher*** | | Student | | Teacher | |
Survey of Classroom Practice Items and Categories (0 to 3 Scale or as indicated)* [Science Items]? = Differ by Site	MN	SD	MN	SD	MN	SD	MN	SD
Homework								
Means for Items			1.71	0.46			1.54	0.44
Minutes 0 = 0, 1 <15, 2 = 15–30, 3 = 31–60, 4 = 61–90, 5 >90 ?	2.10	0.92	2.09	0.80	2.10	1.03	2.02	0.69
Times/week 0 = Never, 1 <1, 2 = 1–2, 3 = 3–4, 4 = 5	2.76	1.08	2.92	0.95	1.87	1.14	1.97	0.83
Count Grades 0 = No, 1 = Usually not, 2 = Usually, 3 = Always	2.36	0.73	2.16	0.88	2.32	0.76	2.26	0.69
Arithmetic computation or procedures	1.81	0.94	2.01	0.87				
Read about science in books, magazines, or articles					1.47	0.94	1.40	0.87
Show required steps	2.07	0.88	2.35	0.77				
Answer questions from science book/worksheet					1.86	0.94	1.52	0.88
Explain reasoning or thinking	1.80	0.91	1.87	0.91			1.42	0.80
Solve science problems requiring computation					1.38	0.97	1.31	0.86
Work on demo presentation or proof	1.87	0.96	1.59	0.95				
Revise and improve students' own work	1.43	1.00	1.25	0.87	1.73	0.98	1.71	0.91
Collect data or information?					1.64	0.98	1.48	0.75
Write a math report or write about science?	0.89	1.02	0.69	0.84	1.37	0.99		
Instructional Activities								
Means for Items			1.45	0.40			1.49	0.34
Watch a teacher demonstration	2.30	0.80	1.93	0.80				

(continued)

254

Table B.1.
(continued)

Survey of Classroom Practice Items and Categories (0 to 3 Scale or as indicated)* [Science Items]? = Differ by Site	Mathematics				Science			
	Student**		Teacher***		Student		Teacher	
	MN	SD	MN	SD	MN	SD	MN	SD
Listen to teacher explain science	1.02	0.93	0.73	0.66	2.19	0.85	1.75	0.78
Read about math or science (non-textbook)	1.35	0.91	1.17	0.72	1.35	0.86	1.55	0.73
Collect information or analyze data	1.42	1.01	1.02	0.92	1.44	0.88	1.44	0.74
Maintain and reflect on portfolio	1.29	0.95	1.72	0.88	1.14	0.96	1.12	0.87
Use hands-on materials or manipulatives?					1.26	0.89	1.39	0.72
Write about science in class					1.61	0.93	1.93	0.90
Engage in problem solving	1.67	0.87	2.09	0.82				
Do lab activity, investigation, or experiment					1.47	0.92	1.25	0.60
Take notes	1.77	0.99	1.32	0.89				
Watch teacher demonstrate an experiment					1.48	0.96	1.92	0.82
Work in pairs or small groups	1.47	0.90	1.88	0.83				
Activity outside the classroom?	0.94	0.92	0.60	0.75	0.77	0.86	0.77	0.75
Use computers or calculators	1.40	0.95	1.64	0.88	0.97	0.94	1.31	0.90
Work individually	1.95	0.90	1.68	0.75	1.86	0.92	1.75	0.72
Take a quiz or test	2.01	0.85	1.38	0.65	1.83	0.93	1.53	0.75
Problem-Solving Activities								
Means for Items			1.61	0.46				
Computations from a text or worksheet	2.03	0.89	1.67	0.85				
Solve word problems from a text or worksheet	1.82	0.83	1.80	0.76				

(continued)

Table B.1.
(continued)

Survey of Classroom Practice Items and Categories (0 to 3 Scale or as indicated)* [Science Items]? = Differ by Site	Mathematics				Science			
	Student**		Teacher***		Student		Teacher	
	MN	SD	MN	SD	MN	SD	MN	SD
Solve novel problems	1.86	0.84	1.25	0.80				
Write an explanation to a problem	1.23	0.90	1.31	0.84				
Apply math to real-world problems	1.46	0.94	1.96	0.77				
Make estimates, predictions, guesses, or hypotheses	1.59	0.88	1.64	0.79				
Analyze data to infer or conclude	1.50	0.92	1.59	0.77				
Laboratory Activities, Investigations, or Experiments								
Follow step-by-step directions					2.10	0.88	1.76	0.56
Use science equipment or measuring tools					1.65	0.93	1.81	0.80
Collect data			0.75	1.67	0.90	1.79	1.82	0.84
Change something in experiment to see what happens					1.33	0.97	1.41	0.77
Design ways to solve a problem					1.35	0.90	1.35	0.90
Make guesses, predictions, or hypotheses					1.70	0.90	1.76	0.81
Make tables, graphs, or charts					1.59	0.89	1.80	0.86
Draw conclusions from science data					1.64	0.92	2.03	0.80
Pairs or Small-Group Activities								
Means for Items			1.50	0.57			1.53	0.54
Talk about solving problems	1.93	0.88	2.09	0.81	1.61	0.89	1.71	0.76

(continued)

256

Table B.1.
(continued)

Survey of Classroom Practice Items and Categories (0 to 3 Scale or as indicated)* [Science Items]? = Differ by Site	Mathematics				Science			
	Student**		Teacher***		Student		Teacher	
	MN	SD	MN	SD	MN	SD	MN	SD
Written assignments from textbook or worksheet	1.88	0.92	1.78	0.85	1.76	0.95	1.65	0.80
Assignment taking longer than one week	1.11	1.04	0.92	0.91	1.52	0.98	1.64	0.89
Work with group to improve written work	1.31	1.01	1.14	1.02	1.30	0.98	1.51	0.85
Review for a test or quiz	2.08	0.96	1.64	0.87	1.89	1.03	1.04	0.80
Write results or conclusions of lab activity					1.56	0.96	1.70	0.92
Hands-on Activities								
Means for Items	1.20	0.97	1.39	0.65			1.65	0.59
Work with hands-on materials to understand concepts	1.47	0.89	1.65	0.97				
Measure objects			1.63	0.97				
Build models or charts	1.24	0.92	1.38	0.80				
Collect data by counting, observing, or surveying	1.28	0.96	1.26	0.78				
Present math information to students	1.27	1.01	1.22	0.81				
Collect Information								
Means for Items					1.76	0.89	1.75	0.83
Ask questions to improve understanding					1.35	0.86	1.67	0.78
Organize and display information in tables or graphs								
Make a prediction based on information or data					1.51	0.88	1.59	0.77
Discuss different conclusions from information/data					1.52	0.90	1.56	0.87

(continued)

Table B.1.
(continued)

Survey of Classroom Practice Items and Categories (0 to 3 Scale or as indicated)* [Science Items]? = Differ by Site	Mathematics				Science			
	Student**		Teacher***		Student		Teacher	
	MN	SD	MN	SD	MN	SD	MN	SD
List pros and cons regarding information					1.27	0.96	1.38	0.86
Reach conclusions or decisions based upon information					1.82	0.95	1.89	0.81
Use Computers								
0 = none, 1 <5/year, 2 <4/month, 3 <4/week, 4 = 4-5/week			1.10	0.50		1.04	0.70	
Learn facts or procedures	1.55	0.95	1.67	0.92	1.43	0.93	1.21	0.95
Use sensors or probes	0.60	0.88	0.32	0.58	0.57	0.85	0.57	0.86
Collect data or info using the Internet	0.98	1.04	0.71	0.76	1.26	1.06	1.07	1.03
Display and analyze information	1.21	0.96	1.16	0.84	1.32	0.90	1.25	1.03
Develop geometric concepts	1.23	0.98	1.20	0.93				
Solve problems using simulations	1.80	1.13	1.37	0.90	1.13	0.99	0.82	0.85
Take a test or a quiz	1.04	1.04	1.13	1.01	1.68	1.14	1.18	1.01
Use individualized instruction or tutorial software					0.95	0.99	0.89	0.96
Use Equipment								
0 = Not Available, 1 = Rarely, 2 = <7/year, 3 = <36/year, 4 = weekly			2.14	0.83			2.02	1.05
Use computer/lab interfacing devices					1.48	1.40	1.33	1.51
Use running water in laboratories (0-4)					1.53	1.39	2.12	1.41

(continued)

258

Table B.1.
(continued)

Survey of Classroom Practice Items and Categories (0 to 3 Scale or as indicated)* [Science Items]? = Differ by Site	Mathematics				Science			
	Student**		Teacher***		Student		Teacher	
	MN	SD	MN	SD	MN	SD	MN	SD
Use electrical outlets (0–4)					1.53	1.37	2.09	1.39
Use other lab equipment (scale, balance) (0–4)					1.92	1.36	2.55	1.09
Use manipulatives (0–4)	1.43	1.27	2.24	1.28				
Use measuring tools (0–4)	1.97	1.23	2.45	1.23				
Use calculators (0–4)	2.08	1.52	2.80	1.39				
Use graphing calculators (0–4)	1.11	1.43	1.16	1.52				
Previous Courses Taken (Secondary Mathematics) N = 1955								
Middle 0-none, 1-1 sem, 2-2 sem, 3-3+	1.77	1.04						
PreAlg AppMath 0-none, 1-1 sem, 2-2 sem, 3-3+	1.55	0.99						
Gen Voc Consumer math 0-none, 1-1 sem, 2-2 sem, 3-3+	0.81	1.00						
Algebra I 0-none, 1-1 sem, 2-2 sem, 3-3+	1.32	1.02						
Integrated 0-none, 1-1 sem, 2-2 sem, 3-3+	0.73	0.97						
Geometry 0-none, 1-1 sem, 2-2 sem, 3-3+	1.03	1.05						
Algebra 2 0-none, 1-1 sem, 2-2 sem, 3-3+	1.01	1.17						
Adv math 0-none, 1-1 sem, 2-2 sem, 3-3+	0.87	1.13						

(continued)

Table B.1.
(continued)

Survey of Classroom Practice Items and Categories (0 to 3 Scale or as indicated)* [Science Items]? = Differ by Site	Mathematics				Science			
	Student**		Teacher***		Student		Teacher	
	MN	SD	MN	SD	MN	SD	MN	SD
Previous Courses Taken (Secondary Science) N = 1322								
General Science 0-none, 1-1 sem, 2-2 sem, 3-3+					1.43	1.02		
Life Science 0-none, 1-1 sem, 2-2 sem, 3-3+					0.96	0.98		
Earth Science 0-none, 1-1 sem, 2-2 sem, 3-3+					1.03	1.01		
Physical Science 0-none, 1-1 sem, 2-2 sem, 3-3+					1.32	1.00		
Biology 0-none, 1-1 sem, 2-2 sem, 3-3+					1.12	1.03		
Chemistry 0-none, 1-1 sem, 2-2 sem, 3-3+					0.76	0.96		
Physics 2 0-none, 1-1 sem, 2-2 sem, 3-3+					0.49	0.85		
Integrated 0-none, 1-1 sem, 2-2 sem, 3-3+					0.78	1.06		

* 0 = None, 1 = <25%, 2 = 25% to 33%, 3 = >33%.
** N Students Mathematics = 2339; N Science = 2221.
*** N Teachers Mathematics = 89; N Science = 74.

Table B.2.
Survey of Classroom Practice Items for Mathematics and Science Teachers

Survey of Classroom Practices Teacher Items (Science Items) N mathematics = 89; N science = 74	Mathematics		Science	
	M	SD	M	SD
School Description				
Class Organization 1 = Departmental, 2 = Subject, 3 = Self-contained, 4 = Team	2.03	1.07	1.94	0.98
Number of Courses Taught (0–7)	2.06	1.44	1.71	1.25
Type Class 0 = Elm., 1 = Middle, 2–5 = Subject (0–5)	1.78	1.77	2.96	2.47
Grade Level 0 = K (0–8)	5.67	1.90	5.98	1.67
Number of Students 0 = <10, 1 <16, 2 <21, 3 <26, 4 <31, 5 >30	3.48	1.24	3.84	1.16
Percent Female (0–9)	4.47	2.03	4.47	1.96
Percent Not Caucasian (0–9)	6.38	3.48	6.92	3.21
Hours instruction per week (0–9 hours)	4.88	1.54	4.53	1.84
Length of Period in min. 0 = NA, 1 <41, 2 <51, 3 <61, 4 <91, 5 <121, 6 = Varies	3.40	1.50	2.93	1.61
Weeks per year 1 <5, 2 <9, 3 <17, 4 <20, 5 <24, 6 <27, 7 <30, 8 <33, 9–36	8.55	1.63	8.01	1.98
Achievement Level of Students 1 = High, 2 = Average, 3 = Low, 4 = Mixed	2.85	0.91	3.04	0.96
Percent LEP 0 = none, 1 <10, 2 <26, 3 <51, 4 >50	0.86	1.10	0.79	1.07
Course Selection 0 = Ability, 1 = LEP, 2 = Teacher, 3 = Parent, 4 = Multiple, 5 = Student	2.38	1.95	2.90	1.85
Most Recent Unit—Percent of Instructional Time in 10% intervals				
Class Periods for Recent Unit (0–9)	1.72	1.38	1.85	1.30
Non instructional (0–9)	1.10	0.87	1.20	0.92
Whole-class lecture or discussion (0–9)	2.25	1.11	2.17	1.09

(continued)

Table B.2.
(continued)

| Survey of Classroom Practices Teacher Items | Mathematics | | Science | |
(Science Items) N mathematics = 89; N science = 74	M	SD	M	SD
Individual student work (0–9)	2.22	1.50	2.12	1.75
Small-group work (0–9)	1.80	1.43	1.31	1.18
Work w/hands-on, manipulatives, lab materials (0–9)			2.32	2.11
Field study or out of class investigation (0–9)	0.49	0.95	0.60	1.33
Student demonstrations or presentations (0–9)	1.13	1.47		
Review or work on homework (0–9)	1.62	1.60	1.11	1.08
Test or quiz (0–9)	1.67	1.75	1.25	0.88
Bring students up to date due to absences or transfers (0–9)	1.04	1.68	0.70	1.25
Assessments— 0 = None, 1 = 1–4/Year, 2 = 1–3/month, 3 = 1–3/week, 4 = 4–5/week	**1.85**	**0.61**	**1.88**	**0.51**
Use objective item assessments	1.63	1.14	1.92	0.74
Use short-answer assessments	2.58	1.01	2.10	0.74
Use extended-answer assessment to explain or justify answers	2.38	1.12	2.23	0.87
Use performance tasks	2.01	1.13	2.21	0.90
Use individual or group presentations	1.56	1.12	1.77	0.99
Use math projects	1.01	0.93	1.32	0.80
Use portfolios	1.10	1.19	1.06	1.08
Use systematic observation of students	2.60	1.19	2.42	1.20

(continued)

Table B.2.
(continued)

| Survey of Classroom Practices Teacher Items | Mathematics | | Science | |
(*Science Items*) N mathematics = 89; N science = 74	M	SD	M	SD
Instructional Influences—0 = NA, 1 = Strong Neg., 2 = Neg., 3 = Little, 4 = Pos., 5 = Strong Pos.	**3.87**	**0.63**	**3.59**	**0.83**
State content standards	4.27	0.88	4.36	0.73
District content standards	4.18	0.97	4.25	0.85
Textbook or instructional materials	3.90	1.10	3.74	1.09
State test	3.83	1.35	3.03	1.86
District test	3.10	1.88	1.93	2.00
National standards	3.97	1.05	3.38	1.67
Experience in pre-service preparation	3.73	1.19	3.68	1.43
Students' special needs	4.24	0.90	3.81	1.33
Parents and community	3.30	1.05	2.99	1.35
Preparation for next grade level	4.45	0.97	4.25	1.13
Instructional Preparation—0 = Not Well, 1 = Somewhat, 2 = Well, 3 = Very Well Prepared	**2.06**	**0.51**	**1.97**	**0.53**
Teach at assigned level	2.61	0.61	2.36	0.73
Use managed cooperative groups	2.10	0.85	2.12	0.71
Implement instruction that meets standards	2.34	0.75	2.23	0.74
Use a variety of assessments	2.18	0.79	1.97	0.90

(continued)

263

Table B.2.
(continued)

Survey of Classroom Practices Teacher Items *(Science Items)* N mathematics = 89; N science = 74	Mathematics		Science	
	M	*SD*	*M*	*SD*
Teach students with physical disabilities	1.17	0.96	1.07	0.99
Help students document and evaluate their own work	1.87	0.80	1.75	0.76
Teach classes with diverse abilities	2.03	0.75	1.85	0.84
Teach students from a variety of cultural backgrounds	2.16	0.81	1.96	0.83
Teach LEP students	1.55	1.08	1.18	0.99
Teach LD students	1.46	0.95	1.30	0.92
Encourage participation of females	2.54	0.62	2.45	0.65
Encourage participation of minorities	2.52	0.66	2.44	0.65
Involve parents in children's education	1.76	0.86	1.62	0.84
Teacher Opinions—0 = Strongly Disagree, 2 = Neutral, 4 = Strongly Agree	**2.33**	**0.44**	**2.14**	**0.43**
All students can learn challenging content	2.77	1.03	2.90	1.03
Enjoy teaching	3.65	0.57	3.47	0.83
Feel supported to try new ideas	3.10	0.89	2.96	1.10
Receive little support from school administration	1.42	1.10	1.37	1.18
Teachers regularly share ideas and materials	2.75	1.13	2.57	1.22
Teachers regularly observe each other teaching	1.26	1.09	1.25	1.06
Have many opportunities to learn new things	2.79	0.94	2.53	1.21
Required to follow rules that conflict with professional judgment	1.53	1.18	1.40	1.19
Teachers actively contribute to curricular decisions	2.07	1.21	1.99	1.07

(continued)

Table B.2.
(continued)

Survey of Classroom Practices Teacher Items (Science Items) N mathematics = 89; N science = 74	Mathematics		Science	
	M	SD	M	SD
Have adequate time to work with peers	1.58	1.27	1.23	0.92
Have adequate materials	2.80	1.07	2.18	1.37
Integrate content with other subjects	2.01	0.85		
Teach estimation strategies	2.28	0.74		
Teach problem-solving strategies	2.42	0.67		
Select or adapt instructional materials	2.14	0.81		
Teach with manipulatives	2.08	0.89		
Asking lots of questions improves learning	3.01	0.85		
Repeated practice is more efficient than using applications and simulations	2.23	1.06		
Use calculators only after mastery of basic facts	2.54	1.27		
Learning is better in similar-ability classes	2.23	1.13		
Learn basic skills before solving problems	2.74	1.24		
Include students' conceptions about natural phenomena			1.71	0.79
Use mathematics in science			2.08	0.76
Integrate science with other subjects			2.18	0.78
Manage class using hands-on or lab activities			2.28	0.81
Learning is better in similar-ability classes			2.35	1.13
Learn basic terms and formulas before concepts and principals			2.45	1.09
Lab-based science more effective than non-lab			3.22	0.91

(continued)

Table B.2.
(continued)

Survey of Classroom Practices Teacher Items *(Science Items) N mathematics = 89; N science = 74*	Mathematics		Science	
	M	*SD*	*M*	*SD*
Testing program dictates science I teach			2.42	1.09
Activity-based science not worth time and expense			0.69	0.94
Absenteeism is a problem	1.95	1.49	1.76	1.27
Mobility is a problem	2.14	1.31	1.78	1.18
Professional Development—0 = No Participat, 1 = Little Impact, 2 = Trying to Use, 3 = Changed Teach	**1.34**	**0.62**	**1.31**	**0.79**
Hours focused on math content 0 = none, 1 <6, 2 <16, 3 <36, 4 >35	2.09	1.24	1.72	1.27
Hours focused on methods of teaching 0 = none, 1 <6, 2 <16, 3 <36, 4 >35	2.26	1.10	1.68	1.25
Implementing content standards	1.84	0.94	1.48	1.07
Implementing new curriculum or instructional materials	1.96	0.84	1.62	1.10
New methods of teaching	1.92	0.88	1.61	1.09
In-depth study of math content	1.40	0.96	1.34	1.08
Meeting the needs of all students	1.75	0.95	1.66	0.96
Multiple assessment strategies	1.93	0.99	1.65	1.12
Educational technology	1.71	0.90	1.75	1.03
Network or study group on improving teaching	1.06	1.13	0.93	1.05
Formal portfolio assessment	0.55	0.88	0.84	1.09
Extended institute or professional development	0.90	1.14	0.73	1.14

(continued)

Table B.2.
(continued)

Survey of Classroom Practices Teacher Items (Science Items) N mathematics = 89; N science = 74	Mathematics		Science	
	M	SD	M	SD
Observed other math teachers	0.54	0.91	0.64	0.94
Read or contributed to professional journals	0.86	1.00	1.10	1.09
Formal Course Preparation—Courses 0 = 0, 1 <3, 2 <5, 3 <7, 3 <9, 5 <11, 6 <13, 7 <15, 8 <17, 9 >17	**2.34**	**2.12**	**3.30**	**2.51**
Refresher mathematics courses	1.94	2.35		
Advanced mathematics courses	2.35	2.60		
Mathematics education courses	2.41	2.41		
Biology/Life Science			3.67	2.96
Physics, Chemistry, Physical Science			3.59	2.93
Geology, Astronomy, Earth Science			2.39	2.33
Science Education			3.33	2.96
Teacher Characteristics				
Years Teaching* 0 <1, 1 <3, 2 <6, 3 <9, 4 <12, 5 <15, 6 >15	4.34	1.77	4.06	1.24
Years Teaching at Current School* 0 <1, 1 <3, 2 <6, 3 <9, 4 <12, 5 <15, 6 >15	3.59	1.76	4.04	1.13
Degree** 0 = Elem., 1 = Middle, 2 = Education, 3 = Subject, 4 = Subject & Edu., 5 = Other	0.69	0.90	0.73	0.99
BA Major** 0 = Elem., 1 = Middle, 2 = Education, 3 = Subject, 4 = Subject & Edu., 5 = Other	2.42	2.12	2.32	1.91

(continued)

Table B.2.
(continued)

Survey of Classroom Practices Teacher Items (Science Items) N mathematics = 89; N science = 74	Mathematics		Science	
	M	SD	M	SD
Adv. Major** 0 = Elem., 1 = Middle, 2 = Education, 3 = Subject, 4 = Subject & Edu., 5 = Other	2.68	2.07	2.62	1.94
Certification*** 0 = Temporary, 1 = Elem., 2 = Middle, 3 = Secondary Other, 4 = Secondary	2.80	1.94	3.23	1.76

Note

*0 = less than 1 year; 1 = 1 to 2 years; 2 = 3 to 5 years; 3 = 6 to 8 years; 4 = 9 to 11 years; 5 = 12 to 15 years; 6 = more than 15 years.

**0 = Elementary Education; 1 = Middle School Education; 2 = Mathematics Education; 3 = Mathematics; 4 = Mathematics Education and Mathematics; 5 = Other Disciplines.

***0 = Emergency or Temporary Certifications; 1 = Elementary Grades Certification; 2 = Middle Grades Certification; 3 = Secondary Certification in a field other than Mathematics or Science; 4 = Secondary Mathematics or Science Certification.

Notes

CHAPTER 1. HISTORICAL CONTEXT

1. The Technical Advisory Network (TAN) was comprised of experts representing fields including anthropology, sociology, statistics, educational research, mathematics, and science education in addition to USI site-representatives who assisted us at both planning and research implementation levels.

CHAPTER 2. THE IMPORTANCE OF DISTRICT AND SCHOOL LEADERSHIP

1. Identifiers were used for all participants in our research. The first two letters refer to the participant's city, and the number identifies the individual.

CHAPTER 4. PROFESSIONAL DEVELOPMENT IN SYSTEMIC REFORM

1. For details about *Schools around the World*, visit www.s-a-w.org.

2. District-level experts were asked to determine the level of intensity at which schools in our study were implementing the USI reforms. Although each site used different criteria to determine school reform level designations, each site considered the participation of teachers from a school in USI professional development activities and, to some extent, the principals' support of the reform agenda.

CHAPTER 5. INSTRUCTIONAL PRACTICES IN MATHEMATICS AND SCIENCE CLASSROOMS

1. The Survey of Classroom Practices is part of the Study of the Enacted Curriculum, an instrument designed and developed by a collaborative effort among the Council of Chief State School Officers (CCSSO), National Institute for Science Education (NISE) at the University of Wisconsin-Madison, National Science Foundation (NSF), and participating states. More information can be obtained from the CCSSO Web site: www.ccsso.org.
2. Names of teachers are pseudonyms.

Works Cited

Abu-Lughod, J. L. (1999). *New York, Chicago, Los Angeles: America's global cities.* Minneapolis: University of Minnesota Press.

Adajian, L. S. (1996). Professional communities, teachers supporting teachers. *Mathematics Teacher, 89,* 321–324.

Amatea, E. S., Behar-Horenstein, L. S., and Sherrard, P. E. D. (1996). Creating school change: Discovering a choice of lenses for the school administrator. *Journal of Educational Administration, 34*(3).

American Association for the Advancement of Science. (1993). *Benchmarks for scientific literacy.* New York: Oxford University Press.

American Federation of Teaching. (2001). *Making standards matter: A fifty-state report on efforts to implement a standards-based system.* Washington, DC: American Federation of Teaching.

Anderson, G. M. (1997). At the Mexican border. *America, 176*(15), 3.

Barnett, C. (1998). Mathematics teaching cases as a catalyst for informed strategic inquiry. *Teaching and Teacher Education, 14*(1), 81–93.

Barnett, C., and Friedman, S. (1997). Mathematics case discussions: Nothing is sacred. In E. Fennema and B. Scott-Nelson (Eds.), *Mathematics teachers in transition.* Hillsdale, NJ: Lawrence Erlbaum.

Barr, R., and Dreeben, R. (1983). *How schools work.* Chicago: University of Chicago Press.

Barth, R. (1988). School: A community of leaders. In A. Lieberman (Ed.), *Building a professional culture in schools* (pp. 129–147). New York: Teachers College Press.

271

Blank, R. K., Porter, A., and Smithson, J. (2001). *New tools for analyzing teaching, curriculum, and standards in mathematics and science.* Washington, DC: Council of Chief State School Officers.

Borko, H., Mayfield, V., Marion, S., Flexer, R., and Cumbo, K. (1997). Teachers' developing ideas and practices about mathematics performance assessment: Successes, stumbling blocks, and implications for professional development. *Teaching and Teacher Education, 13*(3), 259–278.

Boyd, V., and Hord, S. (1994). *Principals and the new paradigm: Schools as learning communities.* Paper presented at the Annual Meeting of the American Educational Research Association, New Orleans, LA.

Briars, D. J. (1999). Curriculum and systemic math reform. *The Education Digest, 64*(7), 22–28.

Bryk, A. S., Sebring, P. B., Derbow, D., Rollow, S., and Easton, J. Q. (1998). *Charting Chicago school reform: Democratic localism as a lever for change.* Boulder, CO: Westview Press.

Cairney, T. H. (2000). Beyond the classroom walls: The rediscovery of the family and community as partners in education. *Educational Review, 52*(2), 163–174.

Calhoun, D., Bohlin, C., Bohlin, R., and Tracz, S. (1997). The mathematics reform movement: Assessing the degree of reform in secondary mathematics classrooms. Paper presented at the annual meeting of the American Educational Research Association, Chicago, IL.

Campbell, P. F., and White, D. Y. (1997). Project impact: Influencing and supporting teacher change in predominantly minority schools. In E. Fennema and B. S. Nelson (Eds.), *Mathematics teachers in transition* (pp. 309–355). Mahwah, NJ: Lawrence Erlbaum.

Carnegie Council on Adolescent Development. (1996). *Turning points: Preparing American youth for the twenty-first century.* New York: Carnegie Corporation.

Carpenter, T. P., Fennema, E., Peterson, P. L., Chiang, C. P., and Loef, M. (1989). Using knowledge of children's mathematical thinking in classroom teaching. An Experimental Study. *American Educational Research Journal 26*(4), 499–453.

Chicago Public Schools Urban Systemic Initiative Year End Report. (1998). Chicago: Chicago Public Schools.

Cobb, P. (1994). Constructivism in mathematics and science education. *Educational Researcher, 23*(7), 4.

Cohen, D. K., and Ball, D. L. (1999). *Instruction, capacity, and improvement* (RR43). Philadelphia: University of Pennsylvania, Consortium for Policy Research in Education.

Cohen, D. K., and Hill, H. C. (2000). Instructional policy and classroom performance: The mathematics reform in California. *Teachers College Record, 102*(2), 294–343.

Cohen, D. K., Raudenbush, S., and Ball, D. L. (2002). Resources, instruction, and research. In F. Mosteller and R. Boruch (Eds.), *Evidence matters: Randomized trials in education research* (pp. 80–119). Washington, DC: Brookings Press.

Cooney, T. J., and Shealy, B. (1997). On understanding the structure of teachers' beliefs and their relationship to change. In E. Fennema and B. S. Nelson

(Eds.), *Mathematics teachers in transition* (pp. 87–110). Mahwah, NJ: Lawrence Erlbaum.

Cooper, H. (2003). Summer learning loss: The problem and some solutions (EDO-PS-03-5). *ERIC Digest.* Champaign, IL: ERIC Clearinghouse on Elementary and Early Childhood Education. Retrieved August 2003, from http://ericeece.org/pubs/digests/2003/cooper03.html.

Corcoran, T., and Goertz, M. (1995). Instructional capacity and high performance standards. *Educational Researcher, 24*(9), 27–29.

Crosswhite, F. J., Dossey, J. A., Swafford, J. O., McKnight, C. C., and Cooney, T. J. (1985). *Second international mathematics study summary report for the United States.* Champaign, IL: Stipes.

Cuban, L. (2001). *Improving urban schools in the twenty-first century: Do's and don't's or advice to true believers and skeptics of whole school reform.* Paper presented at the Symposium on Comprehensive School Reform, Denver, CO.

D'Agostino, J. V. (1996). Authentic instruction and academic achievement in compensatory education classrooms. *Studies in Educational Evaluation, 22*(2), 139–155.

Danzig, A. (1999). How might leadership be taught? The use of story and narrative to teach leadership. *International Journal of Leadership in Education: Theory and Practice, 2,* 117–131.

Danzig, A., and Wright, W. (2002). Narrative and story writing with school leaders: Promises, pitfalls, and pratfalls for learner-centered leadership. Paper presented at the Division A Symposium, *A Continuum of Professional Development for Learner-Centered Leadership,* of the annual meeting of the American Educational Research Association, Chicago, IL.

Datnow, A., and Hubbard, L. (Eds.). (Forthcoming 2004). *Doing gender in policy and practice: Perspectives on single-sex and coeducational schooling.* New York: RoutledgeFalmer Press.

Desimone, L., Porter, A. C., Garet, M. S., Yoon, K. S., and Birman, B. F. (2002). Effects of professional development on teachers' instruction: Results from a three-year longitudinal study. *Educational Evaluation and Policy Analysis, 24*(2), 81–112.

El Paso Collaborative for Academic Excellence. (1998). *Urban systemic initiative year-end report 1997/1998.* Unpublished document, University of Texas at El Paso.

El Paso Independent School District. (2000). *School improvement plans.* Unpublished documents, El Paso, TX.

El Paso Times (1999). Daily newspaper article September 29, 1999.

Franke, M. L., Fennema, E., and Carpenter, T. (1997). Teachers creating change: Evolving beliefs and classroom practices. In E. Fennema and B. S. Nelson (Eds.), *Mathematics teachers in transition* (pp. 255–282). Mahwah, NJ: Lawrence Erlbaum.

Fullan, M. (2001). *The new meaning of educational change* (3rd ed.). New York: Teachers College Press.

Gallagher, J., and Parker, J. (1995). Secondary teacher analysis matrix (STAM). East Lansing, MI: Michigan State University, Department of Teacher Education.

Gamoran, A. (1987). *Instruction and the effects of schooling.* Madison: University of Wisconsin.

Guskey, T. R. (1986). Staff development and the process of teacher change. *Educational Researcher, 15*(2), 5–12.

Hargreaves, A., and Fullan, M. (1998). *What's worth fighting for out there?* New York: Teachers College Press.

Haycock, K. (2001). Closing the achievement gap. *Educational Leadership, 58*(6), 6–11.

Hess, G. A. (1991). *School restructuring: Chicago style.* Newbury Park, CA: Corwin Press.

Heritage Memphis Culture Guide. (1997). Memphis, TN: City of Memphis.

Hewson, P. W., Beeth, M. E., and Thorley, N. R. (1998). Teaching for conceptual change. In K. G. Tobin and B. J. Fraser (Eds.), *International handbook of science education* (pp. 199–218). Dordrecht, Netherlands: Kluwer.

Hord, S. M., and Cowan, D. (1999). Creating learning communities. *Journal of Staff Development, 20*(2), 44–45.

Kahle, J. B. (1998). *Reaching equity in systemic reform: How do we assess progress and problems?* (Research Monograph No. 9). Madison, WI: National Institute for Science Education.

Katzenmeyer W. G. (1999). *The school culture quality survey.* Tampa, FL: The David C. Anchin Center.

Kersaint, G., Borman, K., and Boydston, T. (2001). *The principal's role in supporting professional development.* Paper presented at the meeting of the Systemic Initiative Conference of Key Indicators, Evaluation, Accountability, and Evaluative Studies of Urban School Districts, Tampa, FL.

Kersaint, G., Borman, K., Lee, R., and Boydston, T. (2001). Balancing the contradictions between accountability and systemic reform. *Journal of School Leadership, 11*(3), 217–240.

King, M. B., and Newmann, F. M. (2000). Will teacher learning advance school grades? *Phi Delta Kappan, 81*(8), 576–580.

Kouzes, J., and Posner, B. (1987). *The leadership challenge: How to get extraordinary things done in organizations.* San Francisco: Jossey-Bass.

Krajcik, J., Marx, R., Blumenfeld, P., Soloway, E., and Fishman, B. (2000). *Inquiry based science supported by technology: Achievement among urban middle school students.* Paper presented at the Annual Meeting of the American Association for Research in Education, New Orleans, LA.

Kromrey, J. D., Hines, C. V., Paul, J., and Rosselli, H. (1996). Creating and using a multiparadigmatic knowledge base for restructuring teacher education in special education: Technical and philosophical issues. *Teacher Education in Special Education, 19,* 87–101.

Larson, R., and Csikszentmihalyi, M. (1983). *The experience sampling method.* In H. T. Reis (Ed.), *Naturalistic approaches to studying social instruction.* San Francisco: Jossey-Bass.

Le Tendre, M. J., and Chabran, M. (1998). Title I and mathematics: Making the marriage work. *Journal of Education for Students Placed at Risk, 3*(4), 307–312.

Lee, V. E., Smith, J. B., and Croninger, R. G. (1995). Another look at high school restructuring. More evidence that it improves student achievement and more insight into why. *Issues in Restructuring Schools, 9,* 1–10. Madison, WI: Center on organization and Restructuring of Schools.

Lieberman, A. (1992). *The changing contexts of teaching*. Chicago: University of Chicago Press.

Lieberman, A. (1995). Practices that support teacher development. *Phi Delta Kappan, 76,* 591–596.

Lieberman, A. (1996). Practices that support teacher development: Transforming conceptions of professional learning. In M. W. McLaughlin and I. Oberman (Eds.), *Teachers learning: New policies, new practices.* New York: Teachers College Press.

Little, J. W. (1982). Norms of collegiality and experimentation: Workplace conditions of school success. *American Educational Research Journal, 19*(3), 325–340.

Little, J. W. (1993). Teachers' professional development in a climate of educational reform. *Educational Evaluation and Policy Analysis, 15,* 129–151.

Lortie, D. C. (1975). *Schoolteacher.* Chicago: The University of Chicago Press.

Loucks-Horsley, S., Hewson, P. W., Love, N., and Stiles, K. E. (1998). *Designing professional development for teachers of science and mathematics.* Thousand Oaks, CA: Corwin Press.

Louis, K. S. (1998). *Does professional community affect the classroom? Teachers' work and student experiences in restructuring schools.* Chicago: University of Chicago.

Louis, K. S., Marks, H., and Kruse, S. (1996). Teacher's professional community in restructuring schools. *American Educational Research Journal, 33*(4), 757–798.

McCourt, B., Boydston, T., Borman, K., Kersaint, G., and Lee, R. (2000). *Instructional practices of teachers in four urban systemic initiatives city's school districts.* Paper presented at the meeting of the Systemic Initiative Conference of Key Indicators, Evaluative Studies of Urban School Districts, Tampa, FL.

McKnight, C. C., Crosswhite, F. J., Dossey, J. A., Kifer, E., Swafford, J. O., Travers, K. J., and Cooney, T. J. (1987). *The underachieving curriculum: Assessing U.S. school mathematics from an international perspective.* Champaign, IL: Stipes Publishing Co.

McNeil, L. M. (2000). *Contradictions of school reform: Educational costs of standardized testing.* New York: Routledge.

Memphis City Schools. (1998). *Memphis urban systemic initiative: A report to the National Science Foundation. Urban systemic initiative in science, mathematics, and technology education. Annual Progress Report.* Unpublished documents.

Memphis City Schools. (2000). *School improvement plans.* Unpublished documents, Memphis, TN.

Miami-Dade County Public Schools. (1998). *Miami-Dade County Public Schools Urban Systemic Initiative Year End Report.* Unpublished document, Miami-Dade, FL.

Miami-Dade County Public Schools. (2000). *School Improvement Plans.* Unpublished document, Miami-Dade, FL.

Mizell, M. H. (1992). Private foundations: What is their role in improving the education of disadvantaged youth? In H. Johnston and K. Borman (Eds.), *Effective schooling for economically disadvantaged students: School-based strategies for diverse student populations* (pp. 49–61). Norwood, NJ: Abex Publishing Corporation.

National Council of Teachers of Mathematics. (1991). *Professional standards for teaching mathematics*. Reston, VA.

National Council of Teachers of Mathematics. (2000). *Principles and standards for school mathematics*. Reston, VA.

National Council of Teachers of Mathematics. (1989). *Curriculum and evaluation standards for school mathematics*. Reston, VA.

National Research Council. (1996). *National science education standards*. Washington, DC: National Academy Press.

National Research Council. (2001). *Investigation on the influence of the standards: A framework for research in mathematics, science, and technology education*. Washington, DC: National Academy Press.

National Science Foundation. (2000). *Six critical drivers*. Available at http://www.ehr.nsf.gov/EHR/driver.asp.

A nation at risk: The imperative for educational reform. (1983). Washington, DC: National Commission on Excellence in Education.

Navarro, M. S., and Natalicio, D. S. (1999). Closing the achievement gap in El Paso: A collaboration for K–16 renewal. *Phi Delta Kappan, 80*(8), 597–601.

NEA Foundation for Improvement in Education. (2003). Teachers take charge of their learning: Transforming professional development for student success. Available at http://www.nfie.org/publications/takecharge_full.htm.

Newmann, F. M., and Associates. (1996). *Authentic achievement: Restructuring schools for intellectual quality*. San Francisco: Jossey-Bass.

Newmann, F. M., and Wehlage, G. G. (1995). *Successful school restructuring: A report to the public and educators*. Madison, WI: Center on Organization and Restructuring of School, Wisconsin Center for Educational Research, University of Wisconsin.

O'Day, J., and Smith, M. (1993). Systemic reform and educational opportunity. In S. H. Fuhrman (Ed.), *Designing coherent education policy*. San Francisco: Jossey-Bass.

Okagaki, L., and Frensch, P. A. (1998). Parenting and children's school achievement: A multiethnic perspective. *American Educational Research Journal, 35*(1), 123–144.

Pedhazur, E. J. (1982). *Multiple regression in behavioral research: Explanation and prediction*. New York: Holt, Rhinehart, and Winston.

Peterson, K. D., and Deal, T. E. (1998). How leaders influence the culture of schools. *Educational Leadership, 56*(1), 28–30.

Peterson, P. L., McCarthey, S. J., and Elmore, R. F. (1996). Learning from school restructuring. *American Journal of Education, 33*, 119–153.

Popham, W. J. (2001). Building tests to support instruction and accountability: A guide for policymakers. Commission on Instructionally Supportive Assessment.

Putnam, R., and Borko, H. (2000). What do new views of knowledge and thinking have to say about research on teaching? *Educational Researcher, 21*(1), 4–15.

Resnick, L., and Hall, M. W. (1998). Learning organizations for sustainable education reform. *Daedalus, 127*, 89–118.

Reynolds, A. J. (1991). The middle schooling process: Influence on science and mathematics achievement from the longitudinal study of American youth. *Adolescence, 26*(101), 133–158.

Roeber, E. D. (1999). Standards initiatives and American educational reform. In G. J. Cizek (Ed.), *Handbook of educational policy.* San Diego, CA: Academic Press.

Rosenholtz, S. J. (1989). *Teachers' workplace: The social organization of schools.* New York: Longman.

Ross, S. (2001, August 19). Memphis's school reform models didn't leave to fail. *The Memphis Commercial Appeal.*

Rungeling, B., and Glover, R. W. (1991). Educational restructuring—The process for change? *Urban Education, 25*(4), 415–427.

Saphier, J., and King, M. (1985). Good seeds grow in strong culture. *Educational Leadership, 42*(6), 67–74.

SAS Institute, Inc. (2001). *SAS version 8.2.* Cary, NC: SAS Institute, Inc.

Secretary's Commission on Achieving Necessary Skills. (1991). *What work requires of schools: A SCANS report for America 2000.* Washington, DC.

Senge, P. (1990). *The fifth discipline.* New York: Doubleday.

Senge, P., Cambron-McCabe, N., Lucas, T., Smith, B., Dutton, J., and Kleiner, A. (1994). A fifth discipline. *A fifth discipline fieldbook.* New York: Doubleday Dell.

Senge, P., Cambron-McCabe, N., Lucas, T., Smith, B., Dutton, J., and Kleiner, A. (2000). A fifth discipline resource. Schools that learn. *A fifth discipline fieldbook for educators, parents, and everyone who cares about education* (Rev. ed.). New York: Doubleday Dell.

Senge, P., Roberts, C., Ross, R., Smith, B., Roth, G., and Kleiner, A. (1999). *The dance of change: The challenges of sustaining momentum in learning organizations.* New York: Doubleday.

Shernoff, D. (2001). The experience of student engagement in high school classrooms: A phenomenological perspective. *Department of Education.* Chicago: The University of Chicago.

Shipp, D. (1998). Corporate influence on Chicago school reform. In C. Stone (Ed.), *Changing urban education.* Lawrence, KS: University Press of Kansas.

Simmons, P. E., Emory, A., Carter, T., Coker, T., Finnegan, B., Crockett, D., Richardson, L., Yager, Y., Craven, J., Tillotson, J., Brunkorst, H., Twiest, M., Hossain, K., Gallagher, J., Duggan-Haas, D., Parker, J., Cajas, F., Alshannag, Q., McGlamery, S., Krockover, J., Adams, P., Spector, B., LaPorta, T., James, B., Rearden, K., Lubuda, K. (1999). Beginning teachers: Beliefs and classroom actions. *Journal of Research in Science Teaching 36*(8), 930–954.

Simon, M. A. (1995). Reconstructing mathematics pedagogy from a constructivist perspective. *Journal for Research in Mathematics Education, 26,* 114–145.

Singham, M. (1998). The canary in the mine: The achievement gap between black and white students. *Phi Delta Kappan, 80*(1), 8–15.

Smith, M. S., and O'Day, J. A. (1991). Systemic school reform. In S. Fuhrman and B. Malen (Eds.), *The politics of curriculum and testing.* Bristol, PA: Falmer Press.

Socorro Independent School District. (2000). *School Improvement Plans*. Unpublished documents, El Paso, TX.

Southern Regional Education Board. (1981). *The need for quality*. Atlanta, GA.

Spillane, J. P. (2002). Local theories of teacher change: The pedagogy of district policies and programs. *Teachers College Record, 104*(3), 377–420.

Stein, M. K. (1998). *High performance learning communities. District 2: Report on year one implementation of school learning communities*. Pittsburgh, PA: Pittsburgh University Learning and Research Center (SYN73275) (ERIC Document Reproduction Service No. ED 429 263).

Stein, M. K., Silver, E. A., and Smith, M. S. (1999). Mathematics reform and teacher development: A community of practice perspective. In *The development of professional developers: Learning to assist teachers in new settings in new ways. Harvard Educational Review, 69*(3), 237–269.

Stepick, A., Grenier, G., Castro, M., and Dunn, M. (2003). *This land is our land: Immigrants and power in Miami*. Berkeley: University of California Press.

Stevenson, D. L., Schiller, K. S., and Schneider, B. (1994). Sequences of opportunities for learning. *Sociology of Education, 67*, 184–198.

Stigler, J. W., and Hiebert, J. (1999). *The teaching gap*. New York: The Free Press.

Supovitz, J. A., and Turner, H. M. (2000). The effects of professional development on science teaching practices and classroom culture. *Journal of Research in Science Teaching, 37*(9), 963–980.

Thompson, A. (1992). Teachers' beliefs and conceptions: A synthesis of the research. In D. A. Grouws (Ed.), *Handbook of research on mathematics teaching and learning* (pp. 127–146). New York: Macmillan.

Vaughn, S., Schumm, J. S., and Sinagub, J. (1996). *Focus group interviews in education and psychology*. Thousand Oaks, CA: SAGE Publications.

Westat*McKenzie Consortium. (1998). *The National Science Foundation's Urban Systemic Initiatives (USI) Program: Models of reform in K–12 science and mathematics education*. Washington, DC: Division of Educational System Reform of the National Science Foundation.

What work requires of schools: A SCANS report for America 2000. (1991). Washington, DC: Secretary's Commission on achieving necessary skills.

Williams, B. (1996). *Closing the achievement gap: a vision for changing beliefs and practices*. Alexandria, VA: Association for Supervision and Curriculum Development.

Wilson, S. M., and Berne, J. (1999). Teacher learning and the acquisition of professional knowledge: An examination of research on contemporary professional development. In A. Iran-Nejad and P. D. Pearson (Eds.), *Review of Research in Education* (pp. 173–209). Washington, DC: American Educational Research Association.

Wright, S. (1934). The method of path coefficients. *Annals of Mathematical Statistics, 5*, 161–215.

Yair, G. (2000). Educational battlefields in America: The tug-of-war over students' engagement with instruction. *Sociology of Education, 73*, 247–269.

Ysleta Independent School District. (2000). *School Improvement Plans*. Unpublished documents, El Paso, TX.

Contributors

M. Yvette Baber received her BA in sociology from Baker University in Kansas, her MA in Human Relations Services from Governors State University in Illinois, and her PhD in Applied Anthropology from the University of South Florida. In the years between earning her degrees she worked as a community developer, counselor, and director of various human services programs in the midwest. After a one-year appointment at the University of Memphis as an assistant professor of anthropology, Baber served as a faculty researcher and program coordinator for the NSF grant examining the impact of mathematics and science reform initiatives in four U.S. cities. Baber was the lead researcher in the study of the community context, focusing on the impact of reform on the various community-level stakeholders. She is currently an adjunct faculty member at Penn Valley Community College in Kansas City, Missouri, where she teaches courses in the Social Sciences Division.

Jessica Barber is currently a research analyst with the City of Tampa Department of Strategic Planning and Technology. She has played an active role in a number of research projects, including the NSF study "Assessing the Impact of the Urban Systemic Initiative," the Department of Labor's evaluation of the Youth Opportunity Initiative, Context-Based Research's Post-9/11 Travel Patterns and Attitudes research, and Internet-Based Communities. Her ongoing research on organizational change and the interplay between government and community is being used to encourage cooperative activities within the City of Tampa. Barber is pursuing her PhD in Applied Anthropology at the University of South Florida, with plans to focus on public and private uses and concepts of electronic media.

Kathryn M. Borman is a Professor of Anthropology and Associate Director of the David C. Anchin Center at the University of South Florida. She received her doctorate in the Sociology of Education from the University of Minnesota in 1976. Borman has extensive experience in educational reform and policy as well as evaluation studies. Currently she is working with the American Institutes of Research and NORC on the OERI-funded National Longitudinal Evaluation of Comprehensive School Reform, directing the focus study of forty schools in five districts. She was also the Principal Investigator for the NSF study "Assessing the Impact of the Urban Systemic Initiative," conducted in four urban sites. She served as the editor of the AERA journal *Review of Educational Research* and currently is the editor of the *International Journal of Educational Policy, Research, and Practice.*

Theodore Boydston is a retired science educator from Miami-Dade County Public Schools. He earned his BS from Monmouth College in 1967 and his MA in 1972 from Washington State University. He taught high school physics, biology, and chemistry for eighteen years and was a science chairperson for twelve years in Miami, Florida. In 1987 he left classroom teaching to be a science coordinator in the Miami school district. In 1990 he was appointed a district science supervisor, where he served until he retired in 1998. In 1999 he earned a PhD in science education from Florida State University. He is currently working as a researcher in the David C. Anchin Center.

Bridget Cotner is a research associate with the David C. Anchin Center. She has been part of several evaluation projects as a qualitative researcher, including a National Science Foundation grant, "Assessing the Impact of the National Science Foundation's Urban Systemic Initiative," and currently the "National Longitudinal Evaluation of Comprehensive School Reform" funded by the U.S. Department of Education. She received her BA in Anthropology from Ball State University in 1998 and her MA in Applied Anthropology from the University of South Florida in 2001. She is pursuing her doctorate in the College of Education, University of South Florida.

William Katzenmeyer served as Director of the David C. Anchin Center and Professor in the Department of Measurement and Research from 1994 until his retirement in 2003. Under his leadership the Center expanded from an idea to a research, training, and development program of national significance. After serving as dean of the College of Education from 1978 to 1994, Katzenmeyer was named Dean Emeritus of the College. Prior to coming to USF he served as Associate Dean of the Graduate School of Duke University and as a professor of educational research. While there, he published numerous articles related to measurement and statistical analysis.

Gladis Kersaint, the Co-Principal Investigator for the NSF study "Assessing the Impact of the Urban Systemic Initiative," is an associate professor of mathematics education K–12 at the University of South Florida. She earned her PhD in mathematics education from Illinois State University. Her research interests include factors that influence mathematics teacher education (pre- and in-service), standards-based mathematics teaching, the mathematics teaching and learning of at-risk children, and the use of technology for learning and teaching mathematics.

Jeffrey D. Kromrey is a Professor in the Department of Educational Measurement and Research at the University of South Florida. His specializations are applied statistics and data analysis. His work has been published in *Communications in Statistics*, *Educational and Psychological Measurement*, *Journal of Experimental Education* & *Multivariate Behavioral Research*, *Journal of Educational Measurement, and Educational Researcher*.

Reginald Lee is a research associate at the David C. Anchin Center and a doctoral candidate in the Department of Educational Measurement and Research in the College of Education at the University of South Florida. His main areas of research have focused on minority student misrepresentation in special education and appropriate methodological approaches to evaluating under- and overrepresentation of African American students in specialized programs in public schools.

Kazuaki Uekawa is a sociologist and research analyst at The American Institutes For Research. His specialties are sociology of education, social stratification, and quantitative methods. After graduating from the University of Chicago, he did his postdoctoral training at the David C. Anchin Center at the University of South Florida. He helped design research primarily for the study of student engagement, using Experience Sampling Method. He hopes to continue his study of student engagement, as well as studies in comparative education using Third International Mathematics and Science Study data.

Index